THE MOST INTRIGUING STORY EVER TOLD

*A Summary of Scientific Studies
About How We Got Here*

F. Henry Firsching

University Press of America,® Inc.
Lanham · New York · Oxford

Copyright © 2002 by
University Press of America,® Inc.

4501 Forbes Boulevard, Suite 200
Lanham, Maryland 20706

ISBN 978-0-7618-2191-5 (pbk. : alk. paper)

Contents

PREFACE

I wrote the book because people need to know this information. The book answers age-old questions with realistic answers, not suppositions and mystical longings. The book is a compact source of data and tells the story of how we got here based on a variety of scientific disciplines. All these independent studies come together and present a most remarkable story. Not only is the sequence of events established, but the processes by which the events took place have also been ascertained. Even though each study was entirely independent, each one filled in part of the overall situation.

I personally find this information satisfying, comforting, and in a sense inspiring. This story represents a most remarkable accomplishment for humankind. We should be impressed with the wealth of data as well as awed by the scope and power of the collective information. Many years ago this information appeared to be unattainable. Nevertheless, it has been attained. What a triumph for humankind!

The overall ideas in this book influence my daily thinking and my hopes for the future as well. The information already obtained indicates that further efforts will disclose much more information. Science is the key to understanding our world and exerting some control over it. Science can provide many answers to how humankind can continue to flourish and prosper on the Earth.

ACKNOWLEDGMENTS

I want to thank the many people who helped to bring the book to its final form. Some read the book or parts of it and commented on how it might be improved. Others helped with various aspects of the computer work to bring the book into final form..

I especially want to thank Dr. Kathlyn F. Behm, a librarian at the Southern Illinois University at Edwardsville. She performed a thorough search of the literature on several important aspects of this book. My eldest son, Michael Firsching, was indispensable for performing much of the computer manipulations. My wife, Shirley Firsching, suggested important decisions about how to proceed on several levels.

Dr. Betty White was the most thorough reader and critic about how to proceed. Many of her suggestions were incorporated into the book.. I also wish to thank Sam Smith, former head of the Southern Illinois at Edwardsville news service. He supported my newspaper column and encouraged me to continue writing it. That column proved to be the means for starting this book.

INTRODUCTION

There has been a lot of speculation about how we got here. Many of us have an intense curiosity about our beginnings and that seems to be the driving force for our speculation. But the range of our speculation is now drastically limited by the vast accumulation of knowledge that has recently been collected. The new information clearly shows that many of the old ideas are incorrect. A thorough and comprehensive understanding of how we got here is now available. The complex series of events that made the world, and everything in it, has been unraveled. To really know our history down through the ages gives a feeling of intense satisfaction.

Within this book the series of events are organized into several discreet presentations - astronomy, geology, biology, and anthropology. Even though these topics tend to appear separate, they also tend to intertwine and affect each other. Each area of science conveys part of the overall picture that explains our present situation as well as the way in which the events transpired. Out of necessity, this book begins with some rather complex concepts, because that is how things happened. The format of the book follows from the oldest events up to the present with some speculation about the future added at the end.

The very beginning of the universe is where the book starts. Scientific theory calls this event the BIG BANG. In order to explain this beginning a moderate amount of physics as well as some astronomy must be introduced.

Chapter 1

ASTRONOMY

Our location in the universe is somewhat bizarre. We are living on a relatively small planet, our Earth, which revolves around the Sun. The Sun is a huge, high temperature ball that has a diameter of about 870,000 miles. The distance to the Sun from the Earth is about 93 million miles. The Sun radiates energy that makes life possible on planet Earth. This huge object, our Sun, is just an ordinary star.

The universe is filled with stars. Astronomers now have an understanding of how stars form. Some stars are much bigger, and some stars are considerably smaller than our Sun. Almost all glowing stars are huge balls mostly made of hydrogen and helium. Gravity pulls the atoms together and the size of the star is only limited by the quantity of gas and dust that was available as the star formed. In the early universe there were only two substances available for aggregation into stars - hydrogen and helium. In the more recent past there were additional elements that could be incorporated. Our Sun formed about 4.6 billion years ago and incorporated all the naturally known elements into its structure. Nevertheless, hydrogen and helium represent about 99% of the matter that is present in our Sun, which is not surprising because these two elements are by far the most numerous elements in the universe. Almost all normal stars have a similar composition.

The bigger a star is, the more rapidly it uses up its hydrogen fuel. As the hydrogen fuel becomes depleted the star undergoes a dramatic change. Astronomers call this the death of a star. The death throes of an individual star depend on its size. Big stars exhaust their fuel usually in millions of years. Our Sun is intermediate in size and has a total lifetime of about 10 billion years. Smaller stars use up their hydrogen fuel much more slowly than our Sun. These small stars will continue to glow brightly long after our Sun has been extinguished. Small stars have a lower surface temperature than our Sun. Big stars possess a much higher temperature than medium sized stars like our Sun. Big stars burn their fuel more vigorously than our Sun and have shorter lifetimes. The color of a star is indicative of the surface temperature of that star. Blue-white stars are big, high temperature stars. Yellow colored stars, like our Sun, are medium temperature stars of medium size.

Our Sun is but one star in a vast arrangement of stars called the Milky Way Galaxy. The Milky Way Galaxy is an extremely large assemblage of stars - about 500 billion stars. The overall shape of our galaxy is that of a gigantic pinwheel. The dimensions of the Milky Way Galaxy stagger the imagination. It is about 100,000 light years across the disc. (A light year is the distance that light can travel in a year - about 6 trillion miles. The speed of light is 186,000 miles per second.) Some portions of the Milky Way disc are several 1000 light years thick. The center of the galaxy features a huge bulge. Our Sun is located in the huge disc outside of the central bulge. Our Sun is not unique and would have to be considered a run-of-the-mill star in almost all its properties, including its position within the galaxy

However, the array of planets that rotate about the Sun makes it very special. The interesting question is how did this arrangement come to be. How did our solar system, the Sun and its encircling planets, originate? That requires an examination of how the entire universe came to be. That brings us to the concept called the BIG BANG.

The BIG BANG

When the BIG BANG was proposed in the 1920's it was not the only concept of how the universe began. Several ideas had already been proposed, with the steady state theory being the most popular. However, subsequent observations soon swung the studied opinion of astronomers to the BIG BANG theory. The BIG BANG theory explains so much about our universe. No other theory even comes close to providing such a worthwhile series of explanations.

By the early part of the 1900's considerable progress had been made about understanding the known astronomy. The distance to the nearest star had been determined at 4.3 light years. This is an extremely large distance - about 25 trillion miles away. Traveling at the speed of light, it would take 4.3 years to reach the nearest star. The nearest star is a vast distance away.

Astronomers were trying to understand the organization of the Milky Way galaxy, the vast accumulation of stars that dominates the evening sky. Stars appear as sharp points of light, but other patchy or cloudy looking objects could also be observed in the Milky Way galaxy. These cloudy looking objects were generally labeled nebulae. There were two ideas about these nebulae. One hypothesis suggested that the nebulae were clouds of dust and gas; another theorized that the nebulae might be distant aggregates of stars.

As technology advanced, the telescopes slowly became bigger and better; the number of stars that could be distinguished became more and more numerous. In the early 1920's, astronomers had developed the sophisticated equipment that allowed them to make a detailed examination of the nebulae - the cloud-like objects in the sky. Most of these nebulae are not clumps of gas and dust in our Milky Way Galaxy. Instead, they proved to be gigantic distant clusters of billions of stars. They are independent galaxies. These galaxies are incredibly long distances from our Milky Way

Galaxy. The universe is considerably different than anyone had expected. It is fantastically bigger!

The realization that our solar system is part of a galaxy and that there were innumerable other galaxies was difficult to comprehend. The universe is gigantically bigger than anticipated. Correspondingly, this means that our solar system is considerably less important in the overall scheme of things. This discovery of galaxies increased the known size of the universe by prodigious amounts.

The large galaxy nearest to the Milky Way Galaxy is the Andromeda nebula, about 2.3 million light years away. The light from the Andromeda nebula travels for 2.3 million years to reach us. Humankind's view of the cosmos has undergone a dramatic change. Distances in millions of light-years, even billions of light-years, have become common terms in astronomy. Our view of the universe expanded enormously.

But a further, somewhat astonishing discovery was made. All the large, distant galaxies are moving away from us. No matter in which direction an astronomer looked, the galaxies recede from us. What was even more perplexing is the fact that the farther a galaxy is from the earth, the more rapidly it recedes. If a galaxy is 5 million light years away, it is moving fairly rapidly away from the earth. If a galaxy is 25 million light years away, it recedes from us more rapidly. If a galaxy is 100 million light years away, it moves away even more rapidly.

You can imagine how intrigued astronomers were with this almost bewildering information. After considerable study, a plausible explanation for these phenomena was developed. The data indicated that all of the matter in the universe started from a common point and was then hurled outward in a stupendous blast. This almost inconceivable explosion, the BIG BANG, blasted away all the matter that now comprises our entire universe. The BIG BANG would explain why all these galaxies are moving away from us in every possible direction.

One way of trying to visualize this expansion of the universe is to compare the universe to a raisin cake batter. Picture yourself on one of the raisins. It doesn't matter which raisin you chose. As this batter gets heated in an oven, all the raisins in the batter are moving away from you. The nearest raisins are moving slowly, the more distant raisins are moving a bit faster, and the really distant raisins are moving the most rapidly of all. You can do the same type of analogy with the surface of a balloon. This is similar to a two dimensional version. There are dots all over the surface of this balloon. Picture yourself sitting on one of these dots. Any dot will be just fine. As the balloon is slowly inflated, the dots start moving away from you. The nearest move slightly, the dots a bit farther away move more rapidly, and the farthest dots are moving away the most rapidly. Such considerations indicate that the BIG BANG could explain what is being observed in our universe. This was the first stage of the BIG BANG theory.

Theoreticians began studying this idea in intimate detail. They realized that the conditions of the BIG BANG would severely restrict the formation of many elements. The extremely high temperatures and drastic conditions of the very early universe would severely restrict which elements could form. Similar conditions have been studied in physic laboratories and confirm that essentially only hydrogen and helium could form.

After further exhaustive study and calculations, covering many years, the abundance of the elements in the cosmos was found to agree with the abundance predicted by the BIG BANG theory. This long and difficult experimental work involved meticulous efforts on the part of many chemists, physicists, as well as astronomers. The present elemental composition of the universe did conform to the restrictions imposed by the BIG BANG theory. This was a significant step supporting the theory.

Theoreticians also concluded there would be a uniform background radio wave emission that completely permeated the universe. This was based on the conditions of the early universe before any atoms were produced. Photons of light would be generated and would permeate the entire universe without being absorbed or reemitted. However, once free atoms of hydrogen and helium could exist then such a random profusion of light was no longer possible. Furthermore, as the universe expanded over billions of years of time, this light gradually was converted to longer and longer wavelengths

But when this background radiation was first proposed, there was no suitable equipment for measuring such low level radiation. About 20 years ago this idea was resurrected. Equipment for making this sensitive measurement became available and the background radiation was found. The wavelengths and intensities of this background radiation are exactly the same as predicted by the BIG BANG theory. It was a triumph! This data strongly reinforced the theory.

There is a personal story about this background radiation that is especially interesting and informative. Two Bell Telephone scientists, Arno A. Penzias and Robert W. Wilson, working in Holmdel, N. J., were investigating the possible use of an unusual radio antenna. They became thwarted by the discovery that a continuous background radio noise permeated the entire universe. Wherever they pointed their radio antenna, north or south, east or west, or even straight up, they got this same signal. They got this signal during the day and during the night. They also got it in winter, summer, spring, and fall. The signal was everywhere. They carefully measured its wavelengths and intensities. But they were baffled about it. In a telephone conversation with a colleague, Penzias was informed that a Dr. P. J. E. Peebles at Princeton, about 20 miles away, knew what this background noise was. When they contacted Dr. Peebles, he informed them that it was the background radiation from the Big Bang. Dr. Peebles was then in the process of building a radio telescope to search for the background radiation of the Big Bang. Now he was surprised to discover that someone else, practically next door, had already found it. Dr. Peebles examined the Penzias and Wilson data and was pleased to find that they had identified the same intensities and wavelengths predicted by the BIG BANG theory. What a round about way to verify the validity of the BIG BANG theory. Penzias and Wilson had found it, even though they were looking for something else.

So the BIG BANG theory had started by explaining receding galaxies and ended up correctly predicting a background emission that no one knew existed. The BIG BANG is considered by scientists to be the best description of the origin of our universe. It has withstood the test of time and some vigorous scrutiny. It provides us with an understanding of the universe that is both amazing and perhaps even comforting for some individuals. Perhaps we now know how it all began.

The Structure of the Universe

The universe encompasses everything that is known to exist. Our particular portion of the universe is the Earth - an almost infinitesimally small fraction of the total. The structure of our Earth is a sphere or globe. Most people probably have a fairly good idea of where they are located, with respect to the world as a whole.

But the Earth is not the complete universe. It doesn't even come close. Our Earth is but one member of a bigger unit - the solar system. The planets of this system are: Mercury, Venus, Earth, Mars, Jupiter, Saturn, Uranus, Neptune, and Pluto. All these planets rotate in the same direction about the big central object, the Sun.

The solar system is not the complete universe either. Our Sun is also part of a much bigger system, a galaxy. Our Sun is but one star among about 500 billion stars that make up the Milky Way Galaxy.

A galaxy is an aggregation of stars and also includes big clouds of gas and dust. A galaxy usually comprises about 100 billion to a trillion stars. Our own Milky Way Galaxy is composed of about 500 billion stars that are somewhat similar to our Sun. The shape of our galaxy is much like a giant rotating pinwheel.

However, the organization of the universe does not end with our own galaxy. Our Milky Way Galaxy is just one of the many galaxies in the entire universe. There are about 50 billion galaxies in the entire universe. Our universe is an enormous volume of space in which galaxies are sprinkled. The vastness of our universe strains the imagination.

To comprehend such a wealth of information takes time. Further reading and just plain pondering are necessary. To achieve a true comprehension does take more than a simple reading of a few paragraphs in this book. Do not become discouraged. The events that made our world possible are intrinsically interesting.

Studying the universe and trying to understand it is not easy. Galaxies spread out in every direction. They go on and on, fading into the far distances of the universe. Roughly a million light years of space lie between most galaxies. Nevertheless, even this vast and diffuse array of galaxies possesses some additional organization.

Galaxies appear to be grouped together into clusters and super-clusters. Our own Milky Way Galaxy is part of a small cluster of galaxies, called the local group. In some parts of the universe there appears to be an absence of galaxies - huge voids that do not seem to contain anything significant. A complete understanding of these huge voids has not been achieved. However there are indications that during the early moments of the birth of the universe, huge bubbles may have formed and these voids may be associated with such events. This means that there is additional organization beyond the galaxy level. Further work should bring about a better understanding of this possible organization. Right now the overall picture is a bit unclear.

But problems exist in actually seeing the universe. The distances are so vast. In addition, all the big distant galaxies are moving away from us. The farther a galaxy is from us, the more rapidly it is receding from us. The universe is expanding continuously.

Galaxies that are many billions of light years away from us present a special problem. Elapsed time becomes super-imposed on our observations. As we gaze at these distant galaxies, we are not seeing them the way they are now but the way they were billions of years ago. We are seeing the light that left them in the distant past, and that light has been traveling for billions of years to reach us. We are looking back in time. If these galaxies still exist, the light they are emitting now will not reach us for billions of years into the future. The concept of immense time spans and immense distances gives some idea of the difficulties encountered in trying to decipher our universe.

The fundamental ideas about our universe have been expanded enormously in the last 75 years. The universe is now known to have gigantic dimensions. But that only makes the universe more fascinating.

The formation of the Sun and the entire solar system occurred at the same time, about 8 billion years after the BIG BANG. About 4.6 billion years ago, there was a huge cloud of gas and dust within the Milky Way Galaxy. A supernova explosion erupted in the vicinity of this cloud, initiating the compression of some of the material within the cloud. Additional material from the cloud got drawn into the accumulated mass. The entire gas cloud started to swirl about the condensed portion - our developing sun. With the passage of time, the cloud flattened into a disc-like structure in which material started to condense to form planets. The various planetary bodies attracted additional mass and finally became the members of the solar system, which includes the Earth.

Just how did the Earth get to be as it is now? The first thing to consider is the composition of the Earth. The crustal material is primarily oxygen and silicon. The core of the Earth is mostly iron and nickel. How did these elements get formed? And last but certainly not the least, how did all the trace elements like gold, silver, thorium, uranium, fluorine, phosphorous, boron, and all the others, get formed. It is an interesting scenario and does involve some physics and astronomy as well.

In order to understand the formation of the Earth, we need to know the origin of the material from which the Earth is made. The processes for synthesizing these elements are complex but still understandable.

When the universe is examined, about 99 percent of all the visible matter in the universe is made up of just two elements: hydrogen and helium. An explanation of the origin of these two elements is relatively simple. Under the conditions of the BIG BANG, essentially only hydrogen and helium formed. A trace of lithium was also produced. The formation of all the other elements found on and in the earth is not explained by the BIG BANG. Some other energetic process or processes had to be required.

Further progress in determining the origin of all the other elements required a massive amount of detailed effort by a great many scientists for a long period of time. One of the key items in unraveling this question was the origin of solar energy. It was determined that stars, like our Sun, are operating a hydrogen fusion process. Hydrogen nuclei, inside a star, are being converted into helium nuclei. The overall nuclear reaction is that four hydrogen nuclei are fused together to form a helium nucleus. In this fusion process, vast amounts of energy are released. A nuclear fusion reaction in our Sun is the ultimate source of our sunlight. Our very existence relies on a nuclear reaction that takes place in the Sun's interior.

The sunlight that strikes the Earth today is the end result of nuclear reactions inside the Sun that took place thousands and even millions of years ago. These nuclear reactions take place deep inside the Sun and the energy released takes a great many years to randomly diffuse to the surface of the Sun and be emitted as sunlight. About 400,000 miles of solar material must be traversed in a random process in order for sunlight to escape. It may appear strange, but the sunshine we enjoy today is really the result of "fossil" nuclear reactions that took place in the Sun a very long time ago.

Scientific information clearly shows why elements more complex than helium did not form readily during the BIG BANG. During the normal operation of our Sun elements more complex than helium do not form.

A hydrogen nucleus is composed of a single proton. Every proton has a single positive charge. A helium nucleus is composed of two protons and two neutrons. Neutrons do not possess a charge. Each proton and each neutron have an approximate mass of one atomic mass unit. By far, the most common form of helium is plus two with an atomic mass of four. The simple designation is helium-4.

But two stable isotopes of helium exist. (An isotope is an atom with a specific set of properties. The "iso" infers the same.) Helium-3 is an isotope of helium, just as helium-4 is. Both isotopes behave chemically as helium. Yet their masses are different. Their physical properties are slightly different because of the different masses. Helium-3 is composed of 1 neutron and two protons. Helium-3 is relatively rare, while helium-4 is overwhelmingly the most abundant isotope of helium.

When 2 helium nuclei slam together under extremely high temperature conditions, we would expect the end product would be beryllium-8. Beryllium-8 is the sum total of the charges and masses of the two helium-4 atoms. But beryllium-8 is fantastically unstable. The beryllium-8 nucleus almost instantly decays back to two helium nuclei. This instability of the beryllium-8 nucleus makes our universe possible. If beryllium-8 were stable, our universe would be drastically different from what it is now. Numerous elements beyond beryllium could have formed during the BIG BANG. But actually they did not form, because of the instability of beryllium-8.

Even though the instability of beryllium-8 is so critical, few people know that it is one of the important reasons our universe is the way it is. The profound instability of beryllium-8 is surprising. An even number of protons and neutrons, among the lighter elements, usually produces atoms that are stable. Beryllium-8 appears to be an anomaly. But there can be no doubt about its importance in shaping our universe.

The Origin of the Elements

Stars, including our Sun, can operate the hydrogen fusion process for billions of years, until the hydrogen becomes depleted in the center portions of the star. Then the star undergoes a cataclysmic change, called the red giant phase. During this stage, the internal structure of the star tends to collapse. Temperatures and pressures go up to about 10 to 20 times higher than they had been under normal hydrogen fusion. Under these extreme conditions, three helium nuclei can fuse to form a carbon nucleus. Carbon-12 nuclei are very stable. Once carbon nuclei are formed, additional fusion reactions are possible. A helium nucleus can fuse with a carbon nucleus to form an oxygen nucleus. As oxygen accumulates, additional reactions are possible. Oxygen plus helium produces neon. Neon plus helium yields magnesium, etc. Even atomic numbered elements, both in mass and charge, result from this reaction with helium nuclei. This sequence can occur up to the formation of iron, atomic number 26.

The helium-4 fusion process accounts for a large number of elements that are multiples of helium-4. These include carbon-12, oxygen-16, neon-20, magnesium-24, and silicon-28, sulfur-32, argon-36, and calcium-40. These elements represent some of the most abundant elements in the universe, exclusive of hydrogen and helium which are considerably more numerous than any other elements. The helium-4 fusion process also accounts for the relative abundances of the even numbered elements up to iron.

The helium fusion process accounts for a fairly large number of elements but does not explain how odd atomic numbered elements can form. Nor does it explain how elements heavier than iron can form.

Iron is the last element that can form by helium fusion and liberate energy. Beyond iron, energy is absorbed. This will cause further shrinkage of the star, raising internal temperatures even higher. Additional nuclear reactions are possible. The increased velocity of the nuclear particles causes some fragmentation of nuclei. Some protons and neutrons are released. Both the neutrons and protons can produce higher mass atoms by combining with existing nuclei. Protons can also yield higher atomic number elements by combining with existing nuclei. These processes are called the proton process and the slow neutron process. All the elements up to bismuth-209 can be formed in this way. However, bismuth-209 is the last of the stable isotopes. There are no stable atoms heavier than bismuth-209.

During the final stages of the red giant phase of a star, very energetic nuclear reactions are occurring. All the elements from carbon to bismuth are included in these reactions. Limited amounts of odd atomic numbered elements are produced. But the most abundant elements are still the even atomic numbered elements.

The synthesis of both thorium and uranium is very difficult to explain. Both of these elements are near the end of the atomic scale and both have only radioactive isotopes. The most stable of these radioisotopes have extremely long half-lifes, in the order of billions of years. (A half-life is the time needed for half of a specific radioisotope sample to undergo a nuclear transformation.)

Between bismuth-209 and thorium-232 there are a variety of very short half-life elements. These short half-life radioisotopes decay quickly after being formed. Some of these half-lifes are measured in minutes or seconds. Beyond bismuth there are no stable isotopes. This zone of instability makes it impossible for thorium and uranium to form by any previously described process. The only possibility for these two elements to be made is to have a very large number of neutrons become available in a fantastically short period of time. Astrophysicists deduced that the only way this can occur is during a supernova explosion. During such a runaway nuclear blast, vast numbers of neutrons become available in an instant. This rapid neutron process provides the only way thorium and uranium can be formed.

A supernova is a titanic blast that takes place in the death throes of a large star. The star tears itself apart and blasts all of the outer layers of the star into the surrounding volume of space. Not only is there an explosion, but there is also an implosion. The center portions of the star can be converted into a neutron star or into a small black hole, depending on the size of the dying star.

The supernova explosion then hurls the thorium, uranium, and all the other elements generated in the red giant star, out into space. Even though the supernova debris is originally traveling at extremely high velocities, it eventually slows down. A variety of processes, that influence these high velocity particles, include the following.

The magnetic flux of the galaxy interacts with the charged particles, and gravity also affects them. They also collide with other matter in the galaxy. The synthesized elements then become part of gas clouds in the galaxy. The material in these gas clouds can condense to become stars and planets, much like our solar system.

This explanation only leaves three elements; lithium, beryllium, and boron unaccounted for. These three elements are heavier than helium but lighter than carbon. They cannot survive stellar cookery. In the interior of a star, the extremely high temperatures and pressures will unravel these three elements. Yet they do exist.

Lithium, beryllium, and boron get formed in a special way. In a sense, it is the aftermath of supernovas. Our atmosphere is bombarded by cosmic rays. These cosmic rays are the bare nuclei of atoms traveling at extremely high velocities. Supernova explosions are one of the major sources of cosmic rays. The debris from supernovas comprise some of the nuclei that come smashing into our atmosphere.

Our atmosphere is primarily nitrogen and oxygen. When the nuclei of nitrogen and oxygen atoms are struck by these energetic cosmic ray particles, pieces of the nuclei are blasted loose. The nuclei of both nitrogen and oxygen atoms become fragmented into smaller units. Some of these fragments then become the nuclei of lithium, beryllium, or boron. This fragmentation process in our atmosphere forms these three elements. Lithium, beryllium, and boron are in very low abundance in our world, because of the limited means of producing them. When the formation of lithium, beryllium, and boron is considered, these three elements fall into a special class occupied only by themselves.

This explains the formation of all the elements. Furthermore, calculations about nuclear reactions, involved in the formation of all of these elements, show that the relative abundance of elements in the universe agrees with the probabilities of the nuclear reactions occurring. Even the relative abundances of all the elements have been explained.

The experimental data clearly shows how these elements get formed. Not only have the processes for forming these elements been discovered, but the relative abundances of these elements have also been confirmed. These massive research efforts have provided insights into how we got here. The stars are no longer remote, unreachable entities. The atoms that make up my being were once an integral part of some star activity.

The formation of the particles that make up our universe, our world, and our own bodies, is clearly understood. Surprisingly, the atoms that make up our Earth and us are mostly the residues of exploded stars. We are all made of stardust! The idea that human beings are made of stardust seems to be absurd when it is first heard. Yet the scientific evidence clearly shows that the atoms, that comprise the Earth and our bodies as well, had to originate inside dying stars.

The scientific findings about the origin of the elements became very significant to me as an individual person. My view of the universe and myself received a jolting new outlook. Ecology suddenly had a much broader scope than it did before. My thinking was drastically revised about how I viewed the universe and even myself. I had made a star connection. It is a marvelous idea.

When I was a young man I used to look up at the stars in the evening sky and wonder why they were up there. The stars were gorgeous and awesome as well. It all seemed so mysterious and distant. I was bewildered by the vastness. What could it all mean? What could possibly be the purpose of such a vast and impressive array? I really wanted to know more about this entire scenario. I wanted to have some understanding of the overall situation.

Now I gaze up at the stars in the evening sky with a totally different attitude than I had as a teenager. I realize that without stars I could not possibly exist. Most of the atoms that make up my being were made in stars. And the most amazing attitude of all is that I realize that the atoms that make up my being have already been up there, high in the vast outer space of our galaxy. In a sense, I have been up in the heavens at some time in the distant past. My atoms have toured much of the galaxy. And so have yours. Every living creature on earth is made mostly out of stardust. We are all earthlings, made of Earth stuff and the Earth is almost totally made of stardust. The exceptions being hydrogen and helium and even many of these atoms were once part of stars.

The Formation of the Earth

All this information now makes it possible to discuss the formation of our solar system. If gas clouds, containing all the essential elements, are present in the galaxy, there is a distinct possibility that these materials can condense to form a star and accompanying planets. All it would take is a supernova explosion in the immediate vicinity of a gas cloud. This could cause a shock wave that could compress some of the gas and dust and bring about the start of the condensation process.

The formation of stars has been observed in various stages in our own galaxy. No doubt the formation of our sun took place in similar fashion. The gas cloud begins to rotate and flatten out into a disc-like structure as the material in the center begins to accumulate. The central portion of the condensing gas cloud is where most of the mass collects. In surrounding areas, within the disc-like structure, planets can condense. The planets of the solar system all orbit the Sun in the same direction because they all condensed out of the rotating disc-like structure. The position of the planets conforms to a mathematical progression series.

The central object, our Sun, contains practically all of the mass of the solar system. However, the planets retain most of the angular momentum, of the condensed material of the original gas cloud. Such a set of conditions has been found in other star systems as well. Some newly formed stars are found embedded in discs of debris. Planets have been located around various nearby stars, showing that planets are relatively common. The discovery of planets about other stars was anticipated for a long time. Only in the last few years have any planets been found orbiting other stars. The discovery of planets about other stars is confirmation of many ideas that astronomers have been developing. Our Sun and our solar system are not unique arrangements. Similar orientations are routinely formed as stars condense out of gas clouds. Well over 50 planetary systems have been located already.

The Moon

We still have to account for the tilt of the Earth's axis as well as the earth's moon. Both of these aspects of our world are tied together. The moon is an unusual object. When compared to its parent planet, the Earth, our moon is about 1% of the mass of the Earth. That is a much bigger percentage than any other moon in the solar system. How did our moon come to be? Why are no other moons orbiting the Earth? These questions cry for answers. And now we apparently know the answers.

About 40 years ago, there were plenty of suggestions about the formation of the moon. Scientists who studied moon rocks and moon dust brought back by the lunar-landing expeditions compared the validity of these ideas. Unfortunately, the information obtained from the moon landings did not support any of the prevailing ideas. All these ideas were determined to be wrong.

The failure of all the existing hypotheses brought about a new one - the collision idea. This states that shortly after the Earth formed, it was struck by a Mars-sized object, a big planetesimal about 10% of the mass of the Earth. This was the most violent event in the entire history of the Earth. The collision ejected huge amounts of molten material that then went into orbit around the Earth. Eventually this ejected material coalesced to form the moon. Such a scenario can validly explain the composition of the moon.

The collision hypothesis explains almost all the questions concerning the Earth-moon system. The violent collision disrupted the alignment of the Earth's axis with the Sun. The tilt in the Earth's axis, brought about by this collision, produces the seasons. In the winter time the northern hemisphere of the Earth is tilted away from the Sun. In the summer time the northern hemisphere is tilted toward the Sun. This is the main reason for our seasons. Our world would be distinctly different without winter, summer, spring, and fall. So this ancient collision does influence our activities every day.

The effect this ancient collision now has on our daily lives is profound. The clothing we wear on a given day is dictated by this event. In the winter we clothe ourselves with bundles of garments. In the summer we move about with limited clothing. Our houses are built to keep us warm in the cold winter days as well as cool in the hot summer days. To keep comfortable we need furnaces and air-conditioners. On a given day when it is summer in North

America, in places like Argentina or South Africa, wintry conditions prevail. This ancient collision influences our behavior every day of our lives. There is no doubt about it; past events govern how we live today.

When the gigantic collision between the Earth and a large planetesimal occurred, most of Earth's atmosphere boiled away from the heat generated during the collision. Calculations indicate that the Earth's atmosphere was heated to about 16,000 degrees Celsius. Any atmospheric gases that may have accumulated were lost. A comparison between the atmospheres of Venus and the Earth show a drastic difference, even though Venus and the Earth are about the size. Venus has an atmosphere, mostly composed of carbon dioxide, that is about 90 times as dense as Earth's atmosphere. This information provides a further reinforcement for the collision hypothesis.

Our present hydrosphere (the oceans, lakes, rivers, etc.) as well as our atmosphere were accumulated later by collisions of the Earth with comets and other objects in the early solar system, as well as by outgassing of rocks in the Earth's interior. The density of the air we breath and the amount of water on our world were significantly altered by the ancient collision. The ramifications of that ancient collision influence all life on the Earth. The collision hypothesis provides a clear explanation for our moon, as well as the tilt of the Earth's axis.

This discussion has given us an opportunity to see how astronomy has influenced our lives and the Earth itself. There can be little doubt that astronomy governs just about everything we do. When the Sun comes up in the early morning, another day begins. When the Sun sets, darkness engulfs us. We are regulated by the relative positions of the Earth and the Sun. The nearest star, our Sun, dominates life on Earth. Now we realize that other stars have also played a most vital role in our very existence. The critical elements of the life system are formed inside dying stars. Before an organism can live, stars must die.

Further suggested reading on astronomy and related topics.

Our Universe, An Armchair Guide, Michael Rowan-Robinson, W. H. Freeman Company, N. Y., 1990. (An overview of astronomy.)

Companion to the Cosmos, John Gribbin, Little Brown and Company, 1996. (An overview of astronomy.)

Universe, William J. Kaufmann, III, W. H. Freeman Company, N. Y., 1985. (An overview of astronomy.)

Edwin Hubble, Dale E. Christianson, Farrar, Strauss and Giroux, N.Y., 1995. (A discussion of developments in the mid-1900's.)

The First Three Minutes, A Modern View of the Origins of the Universe, Steven Weinberg, Basic Books, Inc., N.Y., 1977. (A discussion of the BIG BANG theory and how the universe originated.)

The Origin of the Chemical Elements, R. J. Tayler, Wykeham Publications, (London) LTD, 1972. (A highly technical discussion of element formation.)

Galaxies, Timothy Ferris, Stewart, Tabori and Chang, Publishers, N.Y., 1982. (A brilliantly illustrated discussion of galaxies.)

The Whole Shebang, A State of the Universe Report, Timothy Ferris, Touchstone, 1998. (An examination of the universe, its formation and future.)

Supernovae and Nucleosynthesis, David Arnett, Princeton Academic Press, Princeton, N.J. 1996. (A complex treatment, numerous equations and graphs supplement the presentation.)

Visions of Heaven, Tom Wilkie & Mark Rosselli, Hodder & Stoughton, London 1998. (A detailed discussion about the pictures taken by the Hubble Space Telescope.)

Chapter 2

GEOLOGY

In a sense the study of geology is a small part of astronomy. The focus is on our planet rather than the entire universe. We are aware that the Earth is part of the solar system, but we need to know as much as possible about our Earth. The more we know and understand about our planet, the better we will be able to live safely and comfortably on it.

Geology is the study of the Earth. Even though the Earth is a relatively small planet, it is our home and is of overwhelming importance to us. The pictures of the Earth taken by astronauts far above the Earth's surface convey this sense of a fragile, lonely wisp of matter in the huge void of space. The Earth is all we've got.

Geologists in studying the Earth have done a fantastic amount of work. The information collected is substantial. We have a very good idea of just how old the Earth is. The dating of the Earth has been a thorn of contention for a long time, but the age of the Earth has been decisively settled. The Earth is 4.6 billion years old. Let's examine how this dating was accomplished.

The Age of the Earth

Knowledge about the age of the Earth developed over a long period of time, using a variety of methods. Bishop Usher made the first dating of the Earth in 1658, when he studied the lineage of individuals mentioned in the Bible. This was his only source of information. He then decided that the world began in 4004 BC. This date remained unchallenged for a long time. But geologists, studying the rocks that make up the Earth's crust, realized that the Earth must be considerably older.

The Grand Canyon presents an outstanding example of this age problem. The Grand Canyon plunges downward to a depth of about one mile. The walls of the canyon are a sequence of sedimentary rock with only the basement rocks unlayered. (Sedimentary rocks are formed in several ways. Most are made up of fragments of other rocks. These fragments are carried into bodies of water and settle out on the bottom. The wind can also deposit tiny rock particles, such as sand grains. Sedimentary rocks include materials that were once in water solution and then precipitate.) Enclosed in these layered sediments are marine fossils, indicating that the rocks were deposited mostly in seawater. Each layer of sedimentary rock is roughly parallel to the Earth's surface, and on both sides of the canyon the rock sequence is the same. The Grand Canyon is cut from solid rock.

In reconstructing the events that made the Grand Canyon possible, the idea of immense time dominates the situation. Originally, about a mile of sedimentary rock had to be deposited in ocean water, and then this rock had to be uplifted about 9,000 feet (its present elevation above sea level). Both during and after this uplift, the canyon had to be cut by running water to a depth of about one mile.

Comparisons with the current rate of sedimentation taking place in today's oceans indicate that the deposition of the rocks in the Grand Canyon took about 100 to 200 million years. In addition, if we measure the rate at which the Colorado River is removing rock from the canyon now, we obtain some idea of how long such erosion process would take. From this information we can calculate the approximate time needed for the entire operation of cutting the Grand Canyon. It took about 5 to 10 million years.

From the Grand Canyon alone, geologists knew that the Earth must be many millions of years old, not just a few thousand years old.

Two other imaginative ideas for dating the earth developed in the 1800s - heat loss of the Earth, and salinity of the oceans (the concentration of dissolved salts). Both of these indicated that the Earth was at least millions of years old.

The heat loss method assumed that the Earth had once been completely molten. This was established by studying the way earthquakes propagate various waves that radiate in all directions. These earthquake waves move at different velocities when moving through different density material. The earthquake waves are also deflected or scattered at the interface of different density zones within the earth. A study of how these earthquake waves are deflected and scattered disclosed the internal structure of the Earth.

Earthquake waves indicate that the Earth is separated into internal zones. This suggests that the Earth was once molten, permitting the various forms of matter that make up the Earth to separate into zones according to their densities. For example: the core of the Earth is mostly iron and nickel, much denser material than the crust, which is mostly oxygen and silicon. Since that molten period, the Earth has been slowly cooling. Calculation of the time needed for this cooling to take place indicated that the Earth is at least 24 million years old. (However, the more recent discovery of radioactivity, and the heat associated with nuclear decay, now indicates that the Earth is much older than that.)

Salinity of the oceans was also an attempt at dating the Earth, using the time needed to bring the oceans up to their present salt content. As we all know, today's oceans contain a large amount of dissolved material. The rivers of today are continually bringing more dissolved material into the oceans. As rain falls, the rain water is almost completely free of any dissolved minerals. Once the rain strikes the ground there is a distinct possibility that it will encounter soluble minerals like sodium chloride and dissolve some of it. As the water percolates through the soil and then into creeks, and eventually into rivers, the water becomes enriched in minerals.

A calculation was made to determine how long it would take all the oceans, assuming they were originally made only of pure water, to reach the salt content they have today. The volume of water from all of the rivers entering the ocean was considered. The dissolved minerals in the rivers would be emptied into the oceans. A time of about 90 million years is required for the oceans to reach the salinity they have today. .

All of these methods: rock sequence, heat loss, and salinity, indicated that the Earth was indeed very ancient. Yet all failed to determine the true age of the Earth.

Radioactive Dating

With the discovery of radioactivity, entirely new dating methods became possible. In the early 1900s, radioactive dating methods began to be developed. The fundamental idea of such dating is that a radioisotope (any atom that is inherently radioactive), after it undergoes nuclear decay, eventually will be converted into a daughter isotope that is stable. Another important aspect is half-life. For any sample of a specific radioisotope, half of the radioactive atoms will spontaneously decay in a fixed period of time. This is the half-life for that specific radioisotope. Each type of radioisotope has a characteristic half-life. Some radioisotopes have very short half-lifes, while others have extremely long half-lifes. For dating the Earth, only the very long half-life isotopes are useful.

First let us examine just how a specific radioisotope actually functions. Carbon-14 has a half-life of 5,730 years. This means that if a sample contained 1,000 carbon-14 atoms today, 5,730 years from now only 500 atoms of carbon-14 will be present. 11,460 years from now, two half-lifes, only 250 carbon-14 atoms would be left in the sample. For the elapse of one half-life time period, one-half of the radioisotopes present will decay.

This half-life phenomenon is independent of the temperature of the sample. Whether the sample is kept in a tropical jungle or an Antarctic ice sheet does not affect the half-life. Whether the sample is in the form of a gas, or a liquid, or a solid does not affect the half-life of a radioisotope. The radioactive properties of an atom are functions of the nucleus of the atom. Neither the chemical form of a radioisotope nor the environment it is in have any significant effect on the half-life.

In order to do accurate radioactive dating, exacting performance is necessary. Costly equipment must be maintained and calibrated, highly skilled operators must be trained, and exhaustive studies of the specific radioisotope must be made. Only then, can valid and precise dates be obtained. Despite these exacting requirements, numerous accurate radioactive dating methods have been developed.

The validity of these radioactive datings has been demonstrated using ancient samples that are of a known age. From historical records, the precise date of the destruction of Pompeii is known to be 79 AD. A sample of a wooden log from Pompeii was sent to a carbon-14 dating laboratory. The laboratory workers had no idea about the origin of the sample submitted. They performed the carbon-14 analysis and determined the age of the sample. The determined age was within decades of the actual date of the demise of Pompeii. This was a compelling result because the actual date that the tree had been cut was not known. The original tree probably had to be cut a few years before the volcanic eruption. The actual time of the destruction of Pompeii was very close to the carbon-14 dating that was determined, just as had been expected.

Similar checks of the carbon-14 dating method were made using material from Egyptian tombs. From written records, the dates of the various tombs are known fairly well. The relative ages are also known from the sequence of pharaohs. Analysis of the organic material from these tombs produced dates that coincide with the known values from written records. Radioactive dating methods have been shown to be remarkably accurate. The reliability of many radioactive dating methods has been conclusively demonstrated.

Radioactive dating is not limited to carbon-14, although this particular method has been of enormous value to archeology. There are many things that need to be dated. Rock formations, the age of the Earth, the age of the moon, the age of meteorites, etc., go back in time for billions of years. In order to determine elapsed time of billions of years, radioisotopes with half-lifes in the order of billions of years are required. Long half-lifes are needed because short half-life radioisotopes would simply cease to be present in any detectable amounts after a moderate period of time. As an example, let's examine the situation for a radioisotope with a half-life of one year.

Assume there is one gram of the radioisotope present. In one year the amount of the radioisotope would be cut in half. The amount left would be one half gram. Each succeeding year will slash the amount of radioactive atoms in half. The sequence will be one half the first year, then 1/4 the second, 1/8, and with the passage of half-life times the values become 1/16, 1/32, 1/64, 1/128, 1/256, 1/512, and 1/1024. In ten years the amount of the radioisotope present in the sample would be less than one radioisotope out of every thousand originally present. The amount left would be less than a milligram of radioisotope. In twenty years, or twenty half-lifes, the amount of the radioisotope would be less than a microgram - less than a millionth of a gram. Every ten half-lifes the amount of radioisotope present gets decreased by more than a thousand times. When a billion years has gone by, there will be no detectable radioisotope with a half-life of one year left in any sample. Short half-life radioisotopes cannot be used to date objects that are billions of years old, and that includes the Earth.

Carbon-14 would be useless for a determination involving billions of years. In ten half-lifes, 57,300 years, only about one atom of every thousand carbon-14 atoms originally present would still exist in a sample. In 20 half-lifes, 114,600 years only about one atom from every million would still exist in the sample. In the course of billions of years, there would be no detectable carbon-14 present in the sample. Carbon-14 cannot be used to date the age of the Earth.

For making a determination of the age of the Earth, the moon, or meteorites, radioisotopes with half-lifes in the order of billions of years are needed. Eight such isotopes are possible, and five: uranium-238, uranium-235, thorium-232, rubidium-87, and potassium-40 are routinely used. These five radioisotopes have been exhaustively studied, and their decay schemes are known in precise detail.

Often the process of radioactive dating does not involve any measure of the radioactivity of the sample. Instead the analysis of a sample involves a determination of the amount of the original radioisotope that is still present in the sample and the amount of the "daughter" isotope that is present. A mass spectrometer is used for the determinations. The "daughter" isotope is the stable isotope that is eventually produced by the radioactive decay of the "parent" or original radioactive atom. For example: the "parent" uranium-238 decays to the "daughter" isotope, lead-206; but the "parent" uranium-235 decays to the "daughter" lead-207. The sum of the "parent" and the "daughter" isotopes gives the value of the "parent" isotope in the original sample, before any radioactive decay took place.

For example: the half-life of uranium-238 is 4.5 billion years. If 1000 atoms of uranium-238 and 1000 atoms of lead-206 were found in the sample (without any interference from natural lead in the original sample), that sample would be 4.5 billion years old. The original uranium-238 would be the total of the uranium-238 and lead-206 atoms found in the sample. Half of the uranium-238 had decayed to lead-206. One half-life of time had elapsed.

There is another somewhat unusual way to determine the approximate age of the Earth, that involves radioactive dating. In this method the total atmosphere of the earth as well as the overall composition of the Earth are involved. Our atmosphere has an informative composition. The chief components are: nitrogen 78.084%, oxygen 20.946%, argon 0.934%, and neon 0.0018%.

The most unusual feature of this analysis is that argon is almost 1% of the total. In the overall universe neon is considerably more abundant than argon, but in Earth's atmosphere the relative abundance has been drastically reversed. A further complication is that the argon in Earth's atmosphere is almost completely made of one isotope - argon-40. Earth's atmospheric argon is 99.57% argon-40. Once again this does not conform to cosmic abundance. Argon-36 is the most common argon isotope in the universe. The second most abundant isotope is argon-38. An explanation of these discrepancies can provide an approximate age determination of the Earth.

On the Earth the only significant mode of producing argon-40 is by the radioactive decay of potassium-40. Potassium is moderately abundant in the Earth's crust and is composed of several isotopes of potassium. The percentage of potassium-40 in naturally occurring potassium is 0.012%. Potassium-40 forms argon-40, by radioactive decay, with a half-life of 1.3 billion years.

The assumption is made that the argon-40 in the atmosphere is the stable "daughter" of the radioactive potassium-40. During the 4.6 billion years of time since the Earth formed, the argon-40 released by radioactive decay within the Earth could slowly migrate to the surface and be expelled into the atmosphere by volcanic action. So that by sampling the atmosphere now we can get a fairly good idea of how old the Earth really is. Obviously some of the argon-40 was unable to escape from the Earth's interior and that includes most of the radioactive decay that took place in the last 100 million years or so. That is one of the reasons the age determination is approximate. Nevertheless, this determination does give us a general idea of how old the Earth really is.

The total number of argon-40 atoms in the entire atmosphere of the Earth can be calculated because the overall structure of the atmosphere is known very well. The total number of potassium-40 atoms can also be calculated approximately. The concentration of potassium in various rock structures and various lavas has been determined. Estimates of the total quantities of theses lavas and rock structures can be made from densities determined for the Earth's internal structures. It is impossible to sample many areas of the Earth's interior so estimates must be made about their composition. This produces an uncertainty.

If the total number of argon-40 atoms and the total number of potassium-40 atoms were found to be about the same, then the world would be about 1.3 billion years old. The 1.3 billion years is the half-life of potassium-40 and half of the original amount would have been converted to argon-40. However, the actual results show that argon-40 atoms are considerably more numerous than the potassium-40 atoms, indicating that the Earth is considerably older than one half-life. Calculations indicate that the Earth is slightly older than 4 billion years of age.

With the uncertainties in mind, this result provides some reinforcement for age determinations made from rocks. The overall composition of the Earth's atmosphere is another independent method for dating the Earth. This determination tends to confirm the dating of old objects like: primitive Earth rocks, meteorites, and moon rocks. It is always helpful to have two independent procedures give essentially the same value. The idea that things are going correctly is definitely enhanced.

For a typical age determination of a rock that contains some uranium, a determination of the four isotopes; uranium-238, lead-206, uranium-235, and lead-207, would yield two age dates. The radioactive decay of uranium-238 to lead-206 is distinct and independent of the radioactive decay of uranium-235 to lead-207. The half-life of uranium-235 is 0.71 billion years, much shorter than uranium-238's half-life of 4.5 billion years. Yet, the two independent dating determinations can yield essentially the same age for the rock. Confidence in the validity of the age determination of the sample being studied is increased when two different age determinations agree.

Some specific examples of radioactive dating involve moon rocks. Ten samples of basalt, brought back to Earth by the astronauts of the Apollo 17 mission that landed in Mare Serenitatis on the moon, were analyzed. Rubidium-87 dating gave ages that ranged from 3.56 to 3.76 billion years. The average was 3.67 billion years. Argon-40 dates for these samples went from 3.69 to 3.84 billion years, and averaged 3.71 billion years. These are excellent checks for two entirely independent dating procedures. The age for these ten samples is very close to 3.69 billion years, with an uncertainty of about 0.02 billion years. The age for this particular basalt from the Mare Serenitatis region of the moon has been determined by two independent radioactive dating schemes.

Radioactive dating is an extremely valuable method. It has provided an accurate way of determining the timing of many events in the past. We can now look back in time and know precisely when certain events occurred.

There is no doubt in the scientific community that radioactive dating methods give accurate and reliable ages for ancient objects. Critics vociferously say the dating is invalid, but they have no data to support such a claim. Radioactive dating is excellent. The reliability of these methods has been demonstrated over and over again.

There are difficulties with samples that contain thorium, uranium, and natural lead. And an accurate age determination involves corrections for the natural lead that was already present when the original sample formed. However, a discussion of such complications is beyond what this book is about. For anyone interested, there are numerous books on radioactive dating that would provide a thorough explanation.

Radioactive dating techniques have given scientists a reliable figure for the age of the Earth - 4.6 billion years. Even though we cannot find any rocks that old on the Earth, the age of meteorites permitted the determination. Some meteorites are left over from the formation of the solar system. The oldest meteorites ever located are 4.6 billion years old. That must be the time when the solar system was forming. That must be the age of the entire solar system, which also includes the Earth.

The oldest Earth material ever found was 4.2 billion years old. These were zircon crystals incorporated into younger sandstone. The oldest Earth rocks are from northwestern Canada and have an age of 4 billion years.

Scientists are intensely interested in the oldest rocks in the world. The oldest rocks give information about conditions on the Earth shortly after its formation. The most ancient rocks are like small windows to the dim past, through which we can glimpse the events that made our world.

On the Earth there are numerous ways in which rocks get destroyed or reprocessed. Finding material that is around 4 billion years old is indeed very rare. The very old rocks have an extremely small chance to remain unscathed for so long a period of time.

So, these ancient rocks are very difficult to find. The problem is caused by the attrition of rocks that is continually taking place on the Earth. Rocks are being degraded and destroyed by a variety of forces. Volcanic eruptions are probably the most dramatic; but wind, water, ice, and rain also participate. The surf pounding the coastlines of the world also takes it toll.

The movement of continental plates around the globe destroys rocks in a major way. Some of these plates override other plates, and the subducted plate is forced downward into the molten part of the Earth. The rocks melt. This movement of plates goes very slowly, but relentlessly. During the course of millions and billions of years, the quantities of rock that have been obliterated this way must be extremely high.

Throughout the world, rocks are continually being reworked and altered. Rivers and streams tumble rocks and carve out valleys. Such erosion is commonplace. Freezing and thawing also contribute. Such activity makes the chance to find rocks that have survived for about 4 billion years very small.

In Australia, some ancient sandstone has been dated at 3.6 billion years of age. Embedded in these sandstones were some zircon crystals. Apparently, the zircon crystals had been eroded from preexisting rocks and then became mixed in an ancient riverbed along with the sand. These zircon crystals were then isolated from the sandstone and dated. The ages of these zircon crystals were found to be 4.1 and 4.2 billion years old.

These zircon crystals are not truly rocks, but they do represent material that once was part of an ancient rock. Zircons are especially resistant to weathering and degradation. The zircon crystals survived the decomposition of their parent rock, even though the other components of the rock were obliterated.

This means that at least part of the Earth was not molten and would permit the formation of solid rocks 4.2 billion years ago. Perhaps we may never find Earth material that is older than that.

The time sequence of important events has been established and geologists make use of a geologic time table. The first clearly recognized fossil life forms were labeled Cambrian fossils. The Cambrian period covered the time period from about 505 million years ago to 545 million years ago. The older 4 billion years of time is called the pre-Cambrian. However, some very recent fossil finds have produced what are labeled Ediacaran fossils. These earliest macro forms of life occupy a time period of about 600 million years ago to the Cambrian. No hard parts have been found but creatures similar to jellyfish and seapens have been discovered. All earlier forms of life discovered in older rocks have been single celled varieties.

Earth's Internal Heat

As everyone knows there are frequent news reports of volcanic activity in various locations around the world. Molten lava oozes out regularly from the Kilauea volcano in Hawaii. Hot geysers have been surging for decades in Yellowstone National Park. There is no doubt about hot stuff pouring out from the earth's interior. The Earth is much warmer inside than outside. When deep wells are drilled or deep shafts sunk for mines, the temperature of the Earth rises with depth. In a very deep mine in South Africa, a miner broke through a rock barrier and encountered a rush of hot water that scalded the legs of the miners in the area. The interior of the Earth is hot.

There are two fundamental sources for heat in the interior of the Earth: the original heat generated, as the Earth was forming, and continuous radioactive decay. The energy liberated by coalescing bodies of matter in the early periods of our Earth's formation melted all the material. The heat, from the energy of motion in the incoming material as well as from the gravitational compression that accompanied this aggregation, produced an early world that was essentially completely molten. A big part of this type of activity occurred in the first few hundred million years of the Earth's existence. This heating allowed the various components that made up the Earth to separate into layers producing the present structure of the world. The denser material, mostly iron and nickel, settled into the center. The lighter material, mostly oxygen and silicon, became an outer solid crust floating on the molten mantle beneath it. A great deal of this early collision heat has been dissipated during the 4.6 billion years of Earth history.

However, additional heat is still being generated inside the Earth right now. Radioactive isotopes are continuously decaying and releasing energy. The long half-life isotopes of uranium, thorium, potassium, and a few less common radioisotopes, are still undergoing nuclear decay and producing energy. This energy can be released thousands of miles inside the Earth and cannot escape readily. So, the interior of the Earth stays hot.

The warmer material in the Earth's interior tends to migrate slowly toward the Earth's surface. This hot material can break through the crustal surface and produce a volcanic eruption. Geysers and fumaroles are also manifestations of this release of energy from the Earth's interior.

As the heat from the Earth's interior is being brought to the surface, an assortment of energetic phenomena can occur. This movement of warmer material toward the surface produces a force on the Earth's crustal material and can bring about slow movement of the continents. This is labeled continental drift. The motion of segments of the Earth's crust can cause friction, as they rub together, and can produce earthquakes. The internal heat of the Earth is the cause of volcanism, continental drift, and much of the earthquake activity.

Radioactive decay has gone on incessantly for the entire history of the Earth and is still going on. It is a major source of the Earth's internal heat. So, in a strange, roundabout way, many of the major disasters of humankind are due to radioactivity. It is radioactivity that occurred throughout the history of the Earth, millions and even billions of years ago. This heat from radioactive decay keeps the interior of the Earth very hot, and eventually produces the volcanic activity and most of the earthquakes.

When the ground shakes in an earthquake, or when hot lava is ejected from a volcano, that is the way the Earth naturally functions. The internal heat of the Earth is being dissipated, and like it or not, we must learn to live with it.

Continental Drift

This brings us to an important consideration about our Earth, the tectonic plate theory. The idea that the sea floor is spreading was originally proposed about 100 years ago. This theory was based on the shapes of the continents. If the shapes of South

America, Africa, North America, and Europe are cut out of a map they can be moved about like pieces in a jigsaw puzzle. South America and Africa fit together rather nicely. An even better fit occurs if the continental shelves are used, rather than the parts above sea level. South America fits smoothly into the continent of Africa. North America fits almost as well into the European continent.

This remarkable fit of the continental shapes was compelling information, but unfortunately, at that time, there was no known mechanism that could move the continents. About 100 years ago, there appeared to be no logical way that the continents could drift apart. The spreading sea floor theory or continental drift theory did not become widely accepted at that time.

In recent years, mostly in the 1960's, a variety of very decisive information was obtained about continental drift. One compelling source of information was mapping the ocean floor with sonar scanning equipment. In the Atlantic Ocean, there is a prominent ridge right down the middle of the ocean floor. This is now referred to as the Mid-Atlantic Ridge. The term ridge sounds simple but it is a complex feature that consists of two parallel ridges with a rift valley in between. And in the rift valley there is volcanic activity along the complete length of the Atlantic Ocean. There is a succession of smaller ridges, roughly parallel to the main ridge, spreading out with regularity on both sides of the center ridge. Nothing like this had been anticipated.

The participants in this study were astounded. They knew that some new and perhaps even startling information might be obtained, but this was beyond anything they had imagined. The Mid-Atlantic Ridge and the adjacent features indicated decisively that the Atlantic Ocean was slowly and steadily getting wider. The North American continent and the European continent were drifting farther and farther apart.

Small submarines have gone down into the depths of the rift valley and the volcanic activity has been directly observed. There is no question that the sea floor is spreading. The molten material oozes out in the Mid-Atlantic Ridge and slowly increases the ocean floor. The increased dimensions of the ocean cause the continents on both sides to move apart. Everything proceeds at a slow methodical rate. Nevertheless, during periods of millions of years, significant movement takes place.

An aerial magnetic survey was made of the sea floor. An airplane flew back and forth, over the ridge area, recording the magnetic properties of the undersea rocks. An alternating sequence of polarized rock on both sides of the Mid-Atlantic ridge was found. The sequence of polarity is the same on both sides of the ridge, almost a mirror image of each other.

This magnetic alternation was due to the peculiar behavior of the polarity of the Earth's magnetic poles. Occasionally, the polarity of the Earth flips. In an instant, the North Pole of the Earth becomes the South Pole, and vice versa. This has occurred erratically in the past. Usually something in the order of about half a million years passes before a flip occurs in the Earth's magnetic poles. Any rocks cooling from the molten state will reflect the orientation of the Earth's poles as the rock cools. The smaller ridges on both sides of the Mid-Atlantic Ridge formed sequentially, at the same time, and have the same magnetic orientation. The smaller ridges on both sides of the central ridge, where the present day volcanic activity is occurring, exhibit the same magnetic sequences. The magnetic sequencing of the smaller ridges on both sides of the main ridge is a mirror image of each other. The same alternating magnetic sequences found on both sides of the main ridge indicate that the sea floor is forming in the rift valley and slowly moving outward.

Other information comes from the continents themselves. In Antarctica, coal deposits have been found. In order for coal to form, luxuriant forest growth must take place for long periods of time. But in the present climate of Antarctica, trees cannot live. The weather is far too cold. There is no doubt that Antarctica must have been farther north in a much warmer part of the world than its present location today. The Antarctic continent has moved a considerable distance.

Additional information comes from samples taken from the ocean floor and then dated by radioactive methods. The sea floor rocks closest to the Mid-Atlantic ridge are the youngest. Rocks, farther and farther from the ridge, are progressively older. And of even greater significance, the ages on both sides of the Mid-Atlantic ridge coincide. There can be no doubt that the Atlantic Ocean seafloor is spreading and North and South America are slowly moving away from Europe and Africa.

Another surprising discovery about sea floors around the Earth has been made. None of the rocks on any ocean floor have ever been found to be more than 200 million years old. The Earth is over 20 times older than that. What has been happening to the ocean floors?

Logic indicates that if ocean floors are constantly being formed, then old ocean floors must be constantly destroyed. That is what is happening! Mapping of the ocean floors shows that there are subduction zones, areas where the sea floor is plunging under continents. Eventually the plunging sea floor gets melted in the Earth's hot interior. Off the Pacific coasts of Asia and South America are outstanding examples of such activity. Some of the deepest oceans are off the eastern coast of Asia. Mapping of these areas show that the ocean floor dips under the Asian continent. The sites of the subduction zones, where the ocean floor is plunging, are the source of earthquake and volcanic activity. Most of the earthquake and volcanic activity throughout the world occurs at subduction zones.

The Earth's crust consists of six major plates, the American plate, the Pacific plate, the Indian plate, the Eurasian plate, the African plate, and the Antarctic plate. There are a variety of small segments. These six plates represent most of the surface of the Earth. These plates float, or move slowly, on the molten interior of the Earth. In the course of a human lifetime, no obvious movement occurs. However, over periods of millions of years, profound movement takes place. This restless motion of the plates will continue into the future. Our Earth is not "terra firma."

Along the west coast of the United States, the Pacific plate is moving slowly northward with respect to the American plate. In California, the junction of these two plates is called the San Andreas Fault. The plates lock together and the strain gradually builds up, until the rocks rupture, then the plates lurch past one another in an instant. That causes an earthquake!

Earthquakes can occur without volcanism, but volcanoes are always associated with earthquakes. When plates are sliding underneath one another, volcanoes and earthquakes are common. When plates are rubbing against one another, earthquakes without volcanoes can occur.

Of course, earthquakes can occur in places other than plate boundaries, although the vast majority of quakes do occur at the boundaries. The New Madrid Fault in the midwest is a good example of an isolated type of earthquake. This represents a different type of stress on the Earth's crust, other than plate motion. About 15,000 years ago, a huge glacier covered the northern regions of North America. This massive field of ice, over a mile thick, created a downward pressure on the underlying rocks. Once the ice melted away, the underlying rocks tended to spring upward as the huge weight above the crust was removed. The severe New Madrid earthquakes of the early 1800s resulted from this type of stress.

Earthquakes are no longer the mysterious phenomena they once were. Humankind is getting more and more information about why they occur and where they occur. Predicting when they will occur is considerably more difficult. Nevertheless, progress is being made. In the not too distant future we may be able to predict when an earthquake will occur. That will permit all of us to sleep a little better, especially if we live in an earthquake prone area.

Tides and Glaciers

There are a variety of conditions that affect life on the Earth. One of these is the tide. Tides are the high water and low water effects that occur along the coastlines of the oceans. Tides are caused by the gravitational attraction of both the moon and the Sun.

Try to imagine the Earth as a perfect sphere covered with water. Using this simple model helps to understand the tides. The moon's gravity attracts the water that is closest to it, causing the ocean to bulge out slightly. On exactly the opposite side of the Earth is another bulge, caused by the response of the Earth itself to the moon's gravity. So tides appear, 12 hours apart, one facing the moon and the other on the opposite side of the Earth. But there is a further complication.

The Sun also affects the ocean in much the same way as the moon. Even though the Sun is a much more massive object, much heavier than the moon, it is fantastically more distant than the moon. The net effect of the Sun on the ocean tides is about one-third that of the moon. The highest tides occur when the gravity of the Sun and the moon complement each other. The low tides take place when the effects of the gravity of the Sun and the moon tend to cancel each other.

But the actual Earth is not a smooth sphere as the model we have been using. There are continents, islands, ridges on the sea floor, etc. The bulge of water on the ocean's surface does contact these landmasses as the world spins on its axis. Actually the tides do not come in. Instead, the bulge on the ocean is always there and the land turns into the water bulge as the world rotates. The net effect is a surge of the ocean up onto the land, the so-called incoming tide.

Along the coasts of the world, the residents pay close attention to the tides. The tides do effect their immediate environment and in some areas the tides become very high. An understanding of tides is a necessity for people who live near the ocean.

Glaciers are often overlooked as a significant factor in shaping our world. In the not-too-distant past, glaciers were a very important part of the world. Glaciers covered much of the world about 12,000 years ago. Even today, glaciers are still a significant feature of our world. A huge glacier covers almost the entire Antarctic continent. Most of the big island of Greenland in the northern hemisphere is covered by a glacier. There are also a large number of valley glaciers found in the cold northern and southern regions of the world. The present glaciers of the world represent about 2.15% of the world's total water.

About 10 to 15 thousand years ago, the last of four continental ice sheets or continental glaciers covered the northern parts of Europe, Asia, and North America. Even though these continental ice sheets have melted, many of our present-day geographical features are due to glaciations.

Glacier National Park in northern Montana is an outstanding example of a glaciated landscape. The topography is spectacular! Long U-shaped valleys have been scooped out by the slowly moving glacial ice. The shape of some mountains has been modified by once being embedded in the continental ice sheet. Formerly sharp precipitous peaks have been somewhat rounded off by glacial action. By observing the mountains for long distances the extent of the glaciers can be determined.

There are a great many glacial valleys throughout the world. Yosemite in California is probably the most famous glacial valley. It contains amazing landforms and numerous falls created by side streams plunging down the U-shaped canyon gouged out by the glacial ice.

The Great Lakes that lie between Canada and the United States are the most prominent residual glacial features in North America. The Finger Lakes in New York State are another remarkable result of glacial activity.

Many of the hills, the valleys, the lakes, and the rivers in glaciated areas have been changed or molded by the effects of the continental ice sheets. The daily lives of people residing in formerly glaciated areas are affected by the landforms that are now present. The land has been modified by the glaciations. Where they farm, the locations of their homes, and the roads they travel, have all been determined by the glaciations. The continental glaciers are long gone, but in glaciated areas their effects remain to influence almost everything humans do. Events in the past determine much of our present day activities. We may think we are independent of the past, but we are not.

This material about geology brings us up to date somewhat on our present day world. We know when the Earth formed and how it formed. We have looked at some of the things that have shaped our world and still have a significant impact on our daily lives. There are many aspects of geology that have not even been touched, but everything cannot be included in this book. The overall picture has been clarified and that is what I attempted to do when I wrote this book

Suggested reading about geology.

Our Geological Environment, Watkins, Bottino, & Marisawa, W. B. Sanders Company. 1975. (A clear and very helpful presentation about geological phenomena.)

The Planet We Live On, C. S. Hurlbut, Jr., Harry N. Abrams, Inc., Publishers, New York, 1976. (An Illustrated Encyclopedia of the Earth Sciences.)

Earth History, James C. G. Walker, Jones and Bartlett Publishers, Inc., Boston, 1986. (A discussion of the events that formed our Earth the way it is today.)

A Revolution in the Earth Sciences, A. Hallum, Clarendon Press, Oxford, 1973. (The impact of tectonic plate theory on geology.)

Glaciers, Michael Hamburg & Jurg Alean, Cambridge University Press, Cambridge, 1992. (A thorough discussion of glaciers and glaciation around the world.)

The Making of the Earth, Richard Fifield, Basil Blackwell Inc., New York, 1985. (A discussion of a variety of topics by a variety of authors.)

Chapter 3

BIOLOGY

Biology is the study of life. Almost everyone seems to recognize what life is. Many attempts to define life have been made, but a clear comprehensive definition may not be possible. Nevertheless, a dictionary definition follows in order to clarify what is being discussed in this chapter. A dictionary definition of life is, "The property or quality manifested in functions such as: metabolism, growth, response to stimulation, and reproduction, by which living organisms are distinguished from dead organisms or inanimate matter." Perhaps this definition will be of some help.

In this book the emphasis is on the origins of many things, especially ourselves. The subject of biology is a most important aspect of how we got here. We do not know if life on the Earth is unique, or if life is commonplace in the universe. So far, the only place life has been found is here on Earth. Even though there has been a moderate amount of searching, there are no other known forms of life.

Trying to start this chapter with the beginning of life on Earth is awkward. We don't know exactly when life began on Earth, but we do have some strong indications. Every aspect of biology, including its very beginnings, involves the theory of evolution. This chapter begins with the theory of evolution, first proposed by Charles Darwin and Alfred Russell Wallace in 1859.

The Theory of Evolution

The theory of evolution has been, and still is, a source of vehement controversy. Evolution explains so much. It is the guiding principle for understanding biology. From a scientific point of view, billions of facts clearly demonstrate that evolution has happened. There is no doubt about it!

Nevertheless, opinion polls have repeatedly shown that about half of the population of the United States does not believe that evolution is a satisfactory theory. This is most bothersome, because in the scientific community, the theory of evolution is almost unanimously recognized as being correct. Biologists are convinced that the theory of evolution provides a means for understanding a vast assortment of biological information.

The theory of evolution has been with us for about 140 years. Scientists overwhelmingly support it, while some religious oriented individuals denounce it as an attack against god. Let's discuss what the fuss is all about.

The theory of evolution is based on five main statements. Let's take all five and discuss them one at a time. The first two statements are not controversial.

1. Reproduction - Each species begets only its own kind. Horses bring forth only horses; robins produce only robins, etc.

2. Excess - Reproduction exceeds the number of descendants. Trees often produce thousands of seeds, yet only a few take root, and only a small fraction of these ever reach maturity. Rabbits have large litters at frequent intervals, yet only a small number of their offspring ever reach maturity.

3. Variation - Population members are dissimilar. Brothers are not identical, neither are two stalks of corn, nor two trout, etc. Wide differences are often obvious. These variations are usually due to combinations of chromosomes between the parents. This accounts for siblings being both male and female, as well as differing height, weight, coloration, etc. However, there are some

variations that are occasionally more dramatic. An offspring is produced that has characteristics unlike any of his ancestors. This is called a mutant or "sport." A breed of cattle, the polled Hereford, is a good example of such a development. Mating several mutants, cattle without horns that were produced by normal horned Herefords, started this hornless breed. A bull that had a mutation for lack of horns was mated to a cow that had a mutation for lack of horns. Offspring from the matings were also hornless. They were labeled polled Herefords. These mutants were able to reproduce their new characteristics, hornless heads, in their offspring. Such a mutant is now known to be due to a fundamental change in their DNA, the genetic material of reproduction. Mutations are random changes that appear in living organisms and can be inherited by offspring. The genetic change, that produced hornless cattle, was used by cattle breeders to form a new type of Hereford. Some religious people emphatically disagree with this explanation and its implications. They ignore the successful efforts of biologists who have used mutants to produce a variety of agricultural products. These include several breeds of polled cattle.

4. Environmental Selection - Space and resources are limited. Individual organisms compete for food and space. The end result is that individuals with more favorable characteristics for a given environment are better able to compete, and thus leave more descendants than the others. When a wolf pack pursues a herd of deer, the most resourceful wolf often pulls down the weakest deer. In nature, any cripple or misfit is doomed. Only the fit are able to survive. If the fittest survive, they produce more offspring than those who do not survive. Some people disagree with statement 4, even though that is the way nature operates. Under natural conditions the weak and unsuited tend to die, and the more able tend to survive. Predator and prey is a vicious process that has been operating for many hundreds of millions of years. It is one of the driving forces of evolution, and tends to select the fittest for survival.

5. Divergence - Environments vary and organisms respond. Such slow changes repeated through many generations can produce differing organisms.

A dramatic example of such situations exists in certain caves. Blind salamanders and blind fish, that lack pigmentation, are found in some dark caves. The ability to see and skin coloration may be important to survival in normal daylight conditions, but are worthless in the darkness of these caves. These unneeded features have no survival value in dark caves and are slowly lost by creatures isolated in such caves. Such modification has occurred in many caves all over the world. There are many other examples of organisms adapting to differing environments, and all the accumulated facts indicate that organisms do respond to environmental changes. Yet some people object to statement 5, without producing accumulated facts to support the objection. Unless there can be substantial facts produced to negate statement 5, then the accumulated scientific facts must dominant the conclusion.

These five statements are the foundation for the theory of evolution and are the basis for explaining the origin of species.

Support for these ideas comes from the fossil record found in the rocks of the Earth. With the passage of time, the fossils show a progression from simple organisms to more and more complex organisms. The fossil record also shows a progression from only a few types of organisms to a huge assortment of organisms. Change dominates the fossil record. With the passage of time, older life forms have been routinely replaced with newer forms.

The basic idea of evolution is that all forms of life on the present-day world descended from a common ancestor. This would mean that all forms of life on the planet Earth are related to one another to some extent. We are distant cousins of cows, tuna fish, birds, oak trees, ants, and bacteria. Some people find this concept abhorrent. They refuse to believe it! However, biologists and biochemists have clearly demonstrated this relationship in a variety of ways. The most compelling is the DNA evidence. Examination of the DNA of many of the life forms on Earth decisively shows the relationships. There can be no doubt about it. Evolution has happened! The biochemistry of all life forms uses amazingly similar processes. The biochemistry of life is extremely complex. Minor modifications of basic themes do occur, but the overall process remains much the same in whatever organism it is found.

Ever since the theory of evolution was developed there.has been strong opposition to its basic concepts. An example of such derogatory remarks is, "Man did not come from monkeys!" Perhaps you may have noticed that the five main statements of the theory of evolution did not mention this. Nevertheless, the theory accounts for all forms of life, including humans and monkeys. The theory indicates, with help from the fossil record, that humans did not spring forth from monkeys. On the contrary, the evidence shows that humans and monkeys developed from a common ancestor. So far, this common ancestor has not been discovered in the fossil record, although DNA evidence clearly indicates that such a common ancestor once lived. Perhaps the common ancestor will be found some time in the future. (See the section on More Genetic Information About the Origin of Man.)

Unfortunately, about half of the incoming freshmen at colleges and universities have a multitude of misconceptions about evolution. Some of the most common misconceptions are the following. The methods used to determine the ages of fossils and rocks are not accurate. The origin of life by random chance is statistically impossible. Mutations are never beneficial to the organisms.

The first of these misconceptions, about dating fossils and rocks, has been dealt with to some extent in Chapter 2, and concerns the most widely used methods - radioisotope dating. These radioisotope procedures are excellent. Their validity has been shown in numerous ways. There are some more recent dating techniques that do not have the security presented by long use. Yet they also appear to be worthwhile. Questioning the validity of dating methods in one blanket statement is unreasonable. Some are obviously superior to others and the radioisotopes dating methods are very reliable. And most of the dating of fossils and rocks involves radioisotope methods.

Probabilities

The second misconception concerns the origin of life by random chance. Somehow, the strange idea that highly improbable events do not happen gets involved in this conclusion. The idea that improbable events do not happen is wrong, wrong, wrong! Getting struck by lightning is a highly improbable event. Each one of us has about one chance in a million of being killed by a lightning strike each year. Nevertheless, in the United States about 250 people die each year from lightning strikes. The number of United States citizens to die this way is predictable. The improbable events, lightning flash deaths, do happen and they happen year, after year, after year.

Let's delve into this a bit more. Probabilities are a branch of mathematics that deals with the chance that random events will occur. Probabilities can be demonstrated by tossing a coin. Before proceeding further, some conditions must be considered. The coin must be constructed so that the head side has an equal chance of appearing up on a toss, as often as the tail side. Assuming that the coin will not end up on its rim, the coin considered will have but two equal probabilities, landing heads or landing tails.

What are the probabilities that heads will appear on the first toss? Heads is one of the two possibilities. There is half a chance that heads will appear on the first toss. What are the probabilities that two heads will be tossed in succession? There is half a chance that heads will appear on the first toss, and half a chance that heads will appear on the second toss. The results of the first toss do not influence the result of the second toss. Random events have no memory. There is a one-fourth chance that heads will appear twice in two tosses. Stated another way, if two tosses of a coin are made, for four trials, statistically, one of the four trials will produce two heads.

Three heads in a row is less probable than two heads in row. Stated mathematically, it is one-half times one-half times one-half. Meaning that in eight trials, in which a coin is tossed three times, there is one chance that three heads will appear in a row.

There are a great many applications for probabilities. Probabilities can be used to determine the occurrence of events in our daily lives. One of the most common questions asked is, "Are we going to have a boy or a girl?" Fortunately this situation is similar to the toss of a coin. With a few modifications, the results of a coin toss can be applied to human families. Superficially the problem of births many appear identical to a coin toss. There appear to be only two possibilities: a boy and a girl. But boys and girls do not get born at exactly the same rate. Population studies show there are 53 boys born for every 50 girls. If this slight difference is ignored, the outcome for a small family will not be significantly different. The results will be essentially correct, although there is a small bias in favor of boys that is not considered in this treatment.

Assuming a 50-50 probability for boys and girls, then the same calculation as tossing a coin can be used. The probability of having a boy for the first child is one-half. The chance for having two boys in a row is one-fourth. The same figures apply to girls. Half of all couples will have both one boy and one girl for their first two offspring.

What are the chances of having five boys in a row, or five girls? Mathematically that is one-half multiplied by itself five times or one-thirty second. Statistically, one out of every 32 couples that have five children will have five boys in succession. It would be true for five girls. Figures can be calculated for three boys and two girls, etc. Manipulations of this type give a newly married couple an idea of what is most likely to happen. But unfortunately, statistics cannot be applied rigorously to individuals, because each birth is a random event.

When something is improbable, a different approach is used. Suppose there is one chance in a thousand that a given event will occur. That is somewhat improbable. But only if you try it once. Every additional time you try there are additional opportunities for the improbable event to occur. If this improbable event is tried 1,000 times, there is a 63 percent chance that the improbable event will occur. The improbable event has become probable because of the large number of trials. If 10,000 trials are made, then the odds are 19,999 out of 20,000 that the event will occur at least once. The improbable event has become almost a certainty because of so many trials. Improbable events cannot be dismissed as never happening. The number of trials will determine whether the event has a reasonable chance of occurring.

The mathematics of probabilities is very useful. It prepares the way for what is most likely to happen and even alerts us to the chance for unlikely events to occur. In our daily lives, it is most often used in playing games of chance: card games or roll of the dice games. It can answer questions like, "What are the chance that my partner has the ace of spades?" By considering the mathematical probabilities a person can become a better card player, and in the process becomes better able to handle the random events that keep occurring in our lives.

There is also the distinct possibility that several highly improbable steps that lead to life on Earth could have happened just by chance. In terms of molecules, the world is a big place and it has been around for a long, long time. The oceans and other bodies of water cover most of the Earth's surface. The crucial random chemistry that led to life must have taken place in a water environment. And there were hundreds of millions of years available in which the random trials could have taken place. That life is here now, indicates that such a random process did lead to life. No magic is necessary, just fundamental mathematics and probabilities.

Mutations

The third misconception is that mutations are never beneficial to organisms. This misconception fails to consider the way life actually functions. As an example let's examine the way life would be in an antelope herd on the plains area of Africa. The antelope herd consists of several dozen antelopes of the same species. They all appear to be similar in size, shape, and health. Let's assume that a mutation is present in one of the antelope. This is an unfavorable mutation that interferes with the vision of the antelope. The antelope with the unfavorable mutation cannot see as well as the other members of the herd. This antelope cannot see a lion or leopard crouching in the grass as readily as the other members of the herd. This antelope is more likely to be killed than one with better vision. Another antelope suffers from an unfavorable mutation that makes its hearing less acute. This antelope will not be able to hear a predator, like a lion or a leopard, moving in the grass, that the other members of the herd would hear. The antelope with poor hearing will tend to be killed more often than his fellow herd members. The unfavorable mutations tend to disappear from the herd. This is called natural selection, or survival of the fittest.

With favorable mutations, ones that are beneficial, the individual is better equipped to face the rigors of life and thus survive. An antelope with a mutation that produces a slight improvement in vision is able to perceive a predator better than the other members of the herd. This is a slight advantage making this antelope more likely to elude being eaten than the others about him. The more capable antelope will tend to have more offspring because of a longer life. The more favorable features will thus tend to be passed on to the offspring.

The same sort of advantage would occur with an individual antelope that has slightly better hearing than the other members of the herd. This individual is more likely to elude predators and thus more apt to produce more offspring than the members of the herd with poorer hearing. The favorable mutations tend to survive.

Mutations are random events that alter the DNA of an organism. The vast majority of these mutations will be unfavorable. But the unfavorable ones tend to be eliminated by natural processes. The few favorable mutations will tend to be preserved. Over the course of thousands and millions of years, this can have a profound impact on the species.

If two populations of the same species happen to become separated by a mountain range or a deep canyon, then they would be operating in environments that would be somewhat different. Over the passage of time, as mutations took place, the two populations would tend to drift apart in their structure and behavior. Over extremely long periods of time they could develop into different species.

In the Grand Canyon region of Arizona, there are two somewhat similar squirrels on either side of the canyon. They are obviously similar in many ways, but also possess different features that developed because they lived in different environments. The north rim of the Grand Canyon has a similar environment when compared to the south rim. But there is no doubt that there are different environmental conditions in each area. Two different species of squirrel are one of the results of these environmental differences.

The Origins of Life

The really big question now comes up. How did life begin? That is a major obstacle to overcome, but considerable progress has been made in understanding how it could have come about. There are some intriguing indications of how life may have just happened. The Miller-Urey experiment, of about 45 years ago, is an outstanding example of how to achieve some inkling of how life began. Stanley Miller was a graduate student and his mentor was Harold Urey, a Nobel Prize winner. Stanley proposed a study of the early atmosphere of the Earth and how environmental conditions might have changed it. The early atmosphere of the Earth was deduced to be similar to the existing atmospheres of Jupiter and Saturn. Both of these gaseous giant planets have hydrogen, methane, ammonia, and water present in their atmospheres. These four substances were placed in a moderately complicated glass apparatus. The water was heated and then circulated into a large flask where electrical sparking occurred. The sparking was a simulation of lightning flashes in the atmosphere of the Earth.

Dr. Urey was a bit skeptical. He guessed that a very large number of different organic molecules would be produced. There would be so many compounds that any conclusion would be lost in the maze of too much data.

Nevertheless, Stanley Miller did proceed. The mixture was sparked for about a week. The solution inside his apparatus had slowly turned to a brownish color. The contents of the apparatus were then analyzed. A very large assortment of compounds was not produced. Instead the products formed tended to be selective. Prominent among the products were biochemicals. No one had anticipated a selection for biochemicals. (Biochemicals are compounds routinely found in living organisms.) Two of the biochemicals were the amino acids: glycine and alanine.

Glycine is the most common amino acid found in living organisms, and alanine is the second most abundant amino acid found in living organisms. In the Miller-Urey experiment, glycine was the most abundant amino acid found in the products and alanine was the second most abundant. The experiment was a sensation! Biochemists realized that the essential ingredients of the life system might all be synthesized in a similar way.

Numerous additional experiments were tried concerning the formation of key biochemicals under primitive conditions. Practically all the essential biochemicals needed in the life system have been formed under simple conditions in the laboratory. An outstanding example is the biochemical adenine. Adenine is a complex double ring structure molecule that contains five carbon atoms, five nitrogen atoms, and five hydrogen atoms. All fifteen atoms are joined in a very specific orientation. Adenine is one of the base units that comprise DNA. Adenine is also a component of the adenosine triphosphate molecule that is involved in energy transfer within the cells. Adenine is a very essential molecule in the life system.

An attempt to synthesize adenine in a simple way was made. The starting material would be one substance, hydrogen cyanide. Hydrogen cyanide was also a product of the Miller-Urey experiment and is composed of one carbon atom, one nitrogen atom, and one hydrogen atom. Supposedly, five cyanide molecules could combine and produce one adenine molecule. Energy, like ultra-violet light, was used to stimulate the reaction, and to the researcher's astonishment, copious amounts of adenine were produced. The reaction was so easy and so productive that a patent was secured and the procedure became the commercial process for making the adenine compound. The complex molecule, adenine, was very easy to make under simple conditions.

Similar results were often found with a variety of the key biochemicals of the life system. The ease with which such complex biochemicals formed was a surprise. Nevertheless, the production of the building blocks of life does not necessarily mean that the problem of the origin of life had been solved. It had not.

An analogy is the building of a house. All the bricks, boards, nails, doors, windows, etc., are brought to the building site. That is just the necessary beginnings of building a house. Each brick must be placed in a specific way. So must the boards, doors, windows, etc. Everything that comprises a house demands a specific orientation in order to produce a workable house. Normally a team of skilled workmen follows a blueprint to arrange all the house components into their desired locations. A finished house requires a lot of organization.

The same is true for biochemicals. Making all the ingredients of life is essential but it does not produce a living organism until all the ingredients are arranged in an orderly specific way. How did this assembly of molecules in the correct order take place?

You might think there is no way to study this complicated problem. Nevertheless, a great deal of work has been performed on this problem and considerable insights into a solution have been acquired.

I will describe one experiment that I saw performed and in which I participated in a limited way. It involves the production of pseudo-proteins from amino acids in heated water. Three bottles of amino acids were taken from the chemical supply room. A small teaspoon of each of the three amino acids was placed into a beaker of water. The beaker was then heated to boiling on a hot plate. After about an hour of heating, the resulting mixture was taken from the beaker and scanned with a microscope. (That is the part I performed.) Surprisingly, there were numerous cell-like structures scattered throughout the material.

Previous work has shown that the amino acids polymerize into moderately long chains, similar to proteins. They do not have the exact structure of proteins, but they were chains of amino acids. Then these pseudo-proteins organized themselves into cell-like spherical structures. The whole process only took a couple of hours. It was an amazing demonstration. Protein-like material and a cell-like structure formed readily under simple conditions.

We have to realize that life is much more orderly than what took place in that beaker. However, the amazing ease with which these moderately complex structures formed was remarkable. One can visualize that somehow, somewhere, a random process, similar to this, did take place and produce a life-like protein. It certainly removes some of the questions of how life can be assembled spontaneously.

Biochemists believe that in the early stages of the Earth, the lightning flashes in the atmosphere continued to form various products similar to those produced in the Miller-Urey experiment.

These substances collected in the oceans for millions of years and the oceans were thought to be a sort of primordial soup. The ingredients of life were in this soup along with a plethora of other compounds.

The key item needed was a means of replicating a molecular arrangement. Some recent work has focused on this idea of a self-replicating molecular system. Several moderately complex molecular systems have been shown to be self-replicating. Once the original molecule is formed, it is given an opportunity to replicate itself. A solution is prepared that has components for making copies of the original molecules. When the original molecule is placed in this solution, the original molecule acts as a self-catalyst and makes copies of itself. (A catalyst is a substance that promotes a chemical reaction without being destroyed in the promotion.) This is one of the key features of life, self-replication. This rudimentary chemical process has been demonstrated several times. Molecular systems that are self-replicating have been achieved. But such self-replicating systems have not involved the key biochemicals of the life system. That key step is still missing. Nevertheless, progress has been made in producing self-replicating chemical systems.

There is the distinct possibility that life may have just happened when the necessary ingredients merged chemically and were self-replicating. Even though this is a very improbable event, improbable events become more likely if they are tried enough times.

The volcanic zones in the world's oceans impinge on this idea. Today the world oceans have volcanic vents extending for many thousands of miles. The Atlantic Ocean has a volcanic vent that extends from Iceland down to the Antarctic region. Other similar vents are located in the Pacific Ocean and the Indian Ocean. We do not know the extent of volcanic vents in the early Earth, but it may have been somewhat similar. With the world starting out being almost totally molten, the volcanic activity on the floor of the ocean may have been even more extensive than it is now. Unquestionably there was widespread volcanic activity.

At these vents the temperatures are high, over 300 degrees Celsius. The pressures are also high, because over a mile of water is above the volcanic vents. The chemistry under these conditions is only beginning to be studied and some early results indicate that the ingredients of life could undergo considerable changes under these energetic conditions.

Such volcanic vents provide a multitude of places for chemical reactions to occur. These volcanic sites make it possible for an astronomical number of chances for a suitable reaction to take place. Long periods of time were also available. These reactions could occur for hundreds of millions of years. The possibility, even though highly improbable, that one of these reactions would be self-replicating is likely because of the fantastically large number of trials that took place.

So life could have just happened. Nothing in the natural laws of physics or chemistry indicates that it could not have taken place. The chemistry of the origin of life that has been studied indicates that spontaneous generation of life is definitely possible.

This idea has implications for life throughout the universe. The basic units of matter, the elements, are the same throughout the universe. There are so many star systems in so many galaxies that conditions similar to what occurred here on Earth must have occurred in some of the star systems. If so, life could have developed as it did here on Earth. Even though some limited searching has taken place, no definite indicators for life elsewhere have ever been found. Despite this lack of success in locating other life, there is every reason to believe that we are not alone in the entire universe.

Then of course, we have the reality of the situation. Life is present on the Earth. Life arrived somehow. All of the facts on the subject indicate that life just spontaneously happened. Once life got started and could self-replicate in oceans that were a primordial soup of essential nutrients, it was virtually unstoppable. The single molecular arrangement became two, the two became four, the four became eight, etc. Within a few thousand years the oceans would have copies of the original self-replicating molecular arrangement almost everywhere.

That is where evolution could begin its slow but relentless improvement in the life that had spontaneously formed. If one of the self-replicating molecular arrangements underwent a favorable change, it would have an advantage over the others and would eventually replace them. Such one step changes could occur, over and over again, and only the favorable changes would tend to survive. Any unfavorable changes would tend to result in the demise of the organism. Over the course of millions of years a host of favorable attributes could develop and improve the life that existed.

There is additional information about the theory of evolution in the chapter on Archeology and Anthropology involving biochemical molecules. The findings from DNA analysis are especially compelling.

Once life got started it relentlessly persisted, and continually changed and diversified. Let's examine what the study of the earliest life on Earth tells us about our beginnings.

The Earliest Life

We don't know exactly when life began on the Earth, but we do have some strong indications. A reasonable way to start is to look at the search for the earliest life. About fifty years ago geologists were faced with a bewildering dilemma. At that time the oldest organisms known were Cambrian fossils, about 535 million years old. A diverse assortment of these fossils had been located and included corals, pelecypods, trilobites, etc. This was most bothersome. No fossils had been found in any rocks older than the Cambrian rocks, despite some prolonged and diligent searching.

Prominent among these Cambrian fossils were trilobites. The main body of a trilobite is composed of three segments. The exterior coating is made of chitin and is fairly hard. There was a head with something that looked like eyespots, also multiple feet, and a tail-like assembly. Trilobites were very complex creatures. The trilobite must have undergone numerous changes in order to develop such a complex body.

The abrupt appearance of these very sophisticated life forms was exasperating. According to the theory of evolution, there should be much simpler forms of life that preceded these complicated organisms. Where were the ancestors of these creatures?

I vividly recall my geology teacher remarking that these precursors to Cambrian fossils must exist and he emphatically stated that they would be found. Within the next ten years pre-Cambrian fossils had been discovered.

Geologists, including my geology teacher, had realized that some simpler forms of life must have preceded the Cambrian fossils. They guessed that the reason the earlier life forms had not been found might be due to the lack of hard parts, and also that earlier life may have been much smaller.

Geologists started examining ancient rocks very thoroughly with microscopes. They did not aimlessly scan any old rock, but concentrated their efforts on stromatolites. Stromatolites are strange appearing, often grapefruit sized lumps of rocks that came in convoluted layers in some sedimentary deposits. Geologists deduced that stromatolites were brought about by some form of life. Samples of stromatolites and adjacent material were sliced into very thin layers and examined under a microscope, anticipating that single celled life forms might be preserved. After a long and tedious search, in which the semi-transparent slices of rock were microscopically scanned, the first fossil cells were located. This dramatic discovery of an early life form gave impetus to additional searches. Numerous examples of early micro-cellular life forms were eventually located. The oldest of these finds were single-celled fossils that were about 3.5 billion years old. This pushed the earliest life back about three billion years, much closer to the formation of the Earth itself.

Despite these findings, all the experts did not agree that these 3.5 billion year old impressions were true fossils. There was always the possibility that these were random structures that looked like single-celled organisms. Laboratory work had shown that cell-like structures can form without the help of any organisms. The earliest life presented a good case but not a strong case.

However, additional work produced discoveries that strongly reinforced the idea that life existed 3.5 billion years ago. Filaments or strands of cells were located in rocks that are about 3.5 billion years old. The cells look much like beads on a string. Some strands have up to 25 cells in alignment. Numerous examples of these filamentaceous cell structures have been found. The sizes of the cells vary and the cells capping the ends of the strands have different structures than the interior cells. The paleontologists making the discoveries claim that 11 different species of filaments have been found.

These filaments are moderately complex life forms. They seemed to be the beginnings of multicellular organisms. Much simpler life forms must have preceded them. These filament cells infer that life must have been present for a considerable length of time, in order for these moderately complex life forms to evolve. Perhaps life was present on the Earth for several hundred million years before the filament forms of life developed. These filament forms of life have been painstakingly dated at 3.465 billion years. These are clearly recognized as life forms. And they strongly suggest that life began much earlier.

Our Earth is 4.6 billion years old. A study of craters on the moon shows that the moon's surface was subjected to an intense meteorite bombardment up until about 3.8 billion years ago. Unquestionably, the Earth's surface received a similar bombardment because both the Earth and moon were in the same part of the solar system. On the early Earth, meteorite bombardment was intense until about 3.8 billion years ago. Life would have had a difficult time during such bombardment. So a reasonable guess, at the time, would be that life began about 3.7 to 3.8 billion years ago.

That is where things were until additional work was done. Some intriguing analyses have been made of rocks that formed 3.85 billion years ago. These represent the earliest claim of life that has ever been made. The discovery of this evidence for earliest life was made by meticulous examination of sedimentary rocks from the Isua Formation in Greenland. Among the layers of rocks some apatite crystals were found. Apatite is often associated with biological activity. Within the apatite crystals were found inclusions of carbon. These very tiny inclusions were then vaporized in an ion microprobe. (An ion microprobe is a sophisticated modern instrument that measures the amounts of various atoms in a sample.) The isotopic ratio of carbon-12 and carbon-13 were measured in the Isua samples.

The carbon isotope ratio is a means of determining whether a carbon deposit is from biological sources or from inorganic sources. The life system tends to enrich the carbon deposit with carbon-12. The carbon-13 is not totally excluded in the life process, but instead is not as numerous as it would be in an inorganic chemical reaction. Therefore the carbon-12 to carbon-13 ratio becomes a strong indicator for biological origin when the carbon-12 to carbon-13 ratio is very close to the ratio found in living systems.

The carbon-12 to carbon-13 ratio in the Isua rocks is very close to the ratio found in living systems. No other chemical process, except the life process, is known to bring about this ratio. Therefore, the conclusion that life existed in these 3.85 billion year old rocks appears to be valid. This finding is reinforced by the data about filaments or strands of cells. These infer that life began much earlier than 3.5 billion years ago. A long period of time was needed in order for such a relatively complex feature as strands of cells to evolve.

Life must have originated during the intense meteoric bombardment phase and life also persisted until the intense bombardment subsided. The life process apparently survived the bombardment for about 50 million years. This is very interesting information and does give us a better idea of how life began on this planet.

That is where the subject of the earliest life now stands. Apparently life existed 3.85 billion years ago and even that may not be the beginning. There is a distinct possibility that life began before that period. If and when some additional discoveries are made, it may be necessary to place the origin of life even farther back in time. Right now it appears that the first life originated about 750 million years after the primordial Earth formed. If there are any future discoveries of earlier life, it will not change the situation significantly. The earlier that life formed may suggest that life can form more readily than we now expect. The more easily life can form, the better the chances for additional life to occur throughout our universe.

The Progression of Life

The next big development was the eukaryotic cell that has a well defined nucleus. All the earliest cells were prokaryotic organisms that had no nuclei. Eukaryotic cells were an important milestone because all the higher forms of life, including human, are made of eukaryotic cells. The exact time of this development is uncertain but appears to be about two billion years ago. That means that life on the Earth was comprised entirely of prokaryotic cells for about 2 billion years.

The arrival of eukaryotic cells did not mean the demise of prokaryotic cells. On the contrary, numerous example of prokaryotic cells abound in the present day world. Bluegreen algae are an example. However, the development of higher forms of life required eukaryotic cells. So in a sense the arrival of eukaryotic life was a significant advance that made further remarkable developments possible.

The world's atmosphere was also changing. The presence of life was making use of the photosynthesis reaction, in which 6 water molecules, and 6 carbon dioxide molecules react to produce a carbohydrate molecule, and 6 oxygen molecules. About 2 billion years of oxygen production had brought about a significant change in the atmosphere. Hydrogen and oxygen react to form water and the continual renewal of oxygen by way of. the photosynthesis reaction continued to deplete the hydrogen. Eventually, significant concentrations of oxygen became present in the Earth's atmosphere. This change in the atmosphere to oxidizing conditions also occurred about two billion years ago.

Hydrogen gas can escape from the gravity of the Earth, and in about two billion years a great deal of hydrogen was lost into outer space this way. So the atmosphere of the Earth took an extremely long time before it became oxidizing. The fundamental change in the atmosphere made a fundamental change in the way life could function.

Prior to the presence of oxygen in the atmosphere, fermentation was the main process used by organisms to extract energy from their food. But the presence of oxygen permitted the oxidation of food, a much more efficient source of energy for living organisms. In the modern world, higher organisms make use of oxygen in the atmosphere to provide an efficient means of energy extraction from their food sources. A human being, a wolf, a shark, an eagle, etc., all breathe oxygen in order to extract energy as their food is oxidized. The complex life found in today's world can only operate as they do with an efficient energy source, and oxidation does provide such a source.

The next significant discovery was of Edicaran fossils. These are in pre-Cambrian rocks about 600 to 545 million years old. These represent some beautifully preserved fossils. Very delicate impressions of soft-bodied creatures like: jellyfish, worms, seapens, etc., were located. These are the oldest evidence for multi-cellular life forms ever found. Such soft-bodied creatures are seldom preserved as fossils. Nevertheless, the ones that have been found give us a glimpse of the transition from single celled life forms to multi-cellular life forms. It also makes us realize why pre-Cambrian fossils were so difficult to find.

The next important sequence was the Cambrian fossils. They contained hard body parts and were preserved in profusion. Included in these Cambrian fossils were remarkable fossils from the Burgess shale that are nothing short of astonishing.

The Burgess shale is a very small outcrop of rock. It is about eight feet thick, extends across the face of the mountain for about 200 feet, and encompasses an area of a small city block. Yet, this is one of the most important deposits of fossils ever located. The site is close to Banff, Canada and is about 8,000 feet in elevation, up near the crest of the mountain range.

This site has been visited by scientists on numerous occasions. The attraction is the rare fossils that are exquisitely preserved in these shales. So far, about 80,000 fossils have been secured from the Burgess shale. Paleontologists, the people who study fossils, are very excited about these finds.

The Burgess shale is considered to be vitally important for an understanding of the development of complex life forms. The shale deposits are about 530 million years old, placing them in the early Cambrian period of time. This is the time when the first hard parts of living creatures first appeared as fossils. Before that time, no shells, bones, or teeth have ever been found. Many of the Burgess shale fossils do not have hard parts.

The preservation of soft tissue in the Burgess shale was remarkable. The finest details of the anatomy of many creatures were retained intact. Gills, eyes, legs, mouth, etc., have been preserved. By careful study, it is possible to reconstruct many of the small details of the creatures that inhabited the sea floor 530 million years ago. Such exquisite preservation is extremely rare.

These fossils have been studied with painstaking care. The microscope has been used as a tiny layer is lifted to expose the features underneath. The results of this exhaustive study have been truly astounding. Many forms of life in the Burgess shale are totally unlike anything that has ever been seen before.

Currently, most biologists would group all forms of animal life into 17 phylla, or 17 discreet types. But in the Burgess shale there are about 20 additional phylla that now have no living members. Most of these extinct forms are found only in the Burgess shale, although additional sites containing some of these very early life forms have been located. Many of these early extinct forms of life are bizarre. They are unlike anything ever encountered anywhere else.

This discovery of such diverse and different types of life has confounded and excited paleontologists. Their ideas about the progression of life, in the very early stage of complex life forms, had to be drastically reconsidered.

Until the Burgess shale was meticulously deciphered, the assumption had been made that life started out with a small set of life forms and then slowly became more complex. Now we discover that in the very beginning of complex life forms, there was a fantastic assortment of complex life forms. Many of these are so bizarre that they strain the imagination.

The broad assortment of life forms indicates that life mushroomed into a large assortment of possible types in the early Cambrian period. Some of these types were unlike anything alive today. They all could live and manage to survive for a relatively short period of time. But the stress of competition and predation quickly took its toll. The more able life forms continued, while the less able disappeared from the scene

The processes of evolution has whittled these numerous life types down to a relatively few groups. The original experiment in life occurred back in the early stages of the Cambrian period. Some forms were successful, while many others were not. The survival of the fittest eliminated so many types of life. If it were not for the Burgess shale, there would be no record of many of these astounding forms of life

Following the Cambrian period, the successful life forms slowly developed and expanded. There were extinction events that drastically altered the flow of life down through the last 500 million years of time. But the overall effect was that life persisted and gradually become more sophisticated.

The biggest change following the Cambrian period was the development of fish. Life seems to have been almost exclusively in the water realm for a long period of time. But eventually life did emerge from the seas and invade the land. About 400 million years ago, some life forms were present on land - some primitive plants, some scorpion-like animal life, possibly some spiders and insects.

But the amphibians were more dramatic arrivals. Today, frogs, toads, and salamanders, represent amphibians They can move around on land, but must lay their eggs in water, and tend to be associated with water. The continents became the abode of life, as well as the oceans.

The development of reptiles such as: lizards and dinosaurs was the next significant advance. Their eggs did not have to be deposited in water. They became the first big land animals. Flowering plants also developed and began to dominate plant life on the land. The dinosaurs dominated terrestrial life for a long time and near the end of their mastery of the planet a mammal-like creature developed. These early mammals were not very large and were no threat to the dinosaurs. However, an extraterrestrial event would change all that.

About 65 million years ago, the Earth was struck with a large asteroid, probably about 10 miles in diameter. The collision occurred in what is now the Yucatan peninsula in Mexico. The aftermath of this colossal collision caused the demise of much of the world's life forms. All the large terrestrial animals, larger than 50 pounds, were exterminated. The dinosaurs were totally obliterated. This left the Earth's surface free to be exploited by the mammal-like creatures that survived the impact and its after effects.

There has been a great deal of speculation of what brought about the demise of the dinosaurs. However, the information obtained about the asteroid impact successfully accounts for the events that transpired 65 million years ago. No other dinosaur extinction hypothesis is capable of presenting such a satisfactory explanation.

With the dinosaurs gone, the mammals took over the globe rather quickly. The predators like cats and canines flourished. The herbivores like cattle and antelope grazed on the plant world. A rather insignificant form of the mammal, the primate, slowly developed into a much more formidable creature, man.

The speed with which the present day fauna developed was remarkable. The asteroid impact had changed the world in a most dramatic fashion. It made our present world possible. As long as the highly successful dinosaurs existed, the lowly mammal-like creatures would not have much of a chance. The mammals thrived in a world without dinosaurs and proliferated into a vast assortment of animal forms that populate the world of today. And it all happened in just 65 million years.

We have tiny mammals like mice and squirrels, slightly bigger forms like raccoons and ground hogs, and larger forms like lions and horses. Then there are gigantic forms like elephants and whales. Mammal life dominates our world. And the most dominant of all is man.

One important conclusion can be made from all this information. Everything that is alive today is almost 4 billion years old. Once life formed it persisted. The breathe of life, the essence of life has been handed down from individual being to individual being. As an individual we may be 27 years old or 53 years old. But as a life form we are at least 3.85 billion years old or more. Every ant, bird, or bush, is also a descendant of the first living form and they are also almost 4 billion years old. The renewal of life in each generation is one of the remarkable aspects of life. We can be young but still be incredibly old at the same time.

Extraterrestrial Life

Humankind has a fascination with life beyond the Earth. The excitement generated by so-called UFO's is all too common. Many people would like to know if there really is life in outer space, especially intelligent life.

So far, there is very little factual information to support the idea of life elsewhere. Sure, planets have been found around nearby stars, but this does not assure that any life is present. If life is on any of these remote planets, we have not found evidence to support such an idea. Nevertheless, the fact that life did form on planet Earth does indicate that such a series of improbable events are not impossible. A similar scenario could occur on any planet, anywhere, if conditions were similar to what we have here on Earth.

If we invoke statistics, reasoning suggests that given enough trials the emergence of life is possible. This presents the chance to reach some sort of vague conclusion about life and even intelligent life. There are about 50 billion galaxies in the entire universe. Each galaxy is composed of hundreds of billions of stars. In our own galaxy - the Milky Way - we do know that life and intelligent life has formed. We don't know if Earth is the sole abode of life. Perhaps there may be hundreds or thousands of intelligent life forms within our own galaxy. We don't even know if any other life exists within our own galaxy. Nevertheless we do have that one fact - intelligent life is present in our galaxy.

With about 50 billion galaxies in the universe, it seems reasonable that similar situations that occurred here on Earth might be possible. If we assume that only one out of every fifty galaxies produced intelligent life, then a billion intelligent life forms are possible. The chance that intelligent life formed elsewhere appears to be very high.

There is one almost overpowering difficulty about this conclusion. There is no way to determine if it is correct. The nearest big galaxy to the Milky Way is the Andromeda nebula. It is about 2.3 million light years away. That is an extremely awkward position. It is difficult to comprehend how a message can be sent over such an enormous distance. If we somehow received a message from "them", it would take our return message 2.3 million years to get there. If "they" replied, it would take another 2.3 million years to reach us. Communications, if at all possible, would be frustratingly slow. Excruciatingly slow might be a better description. A period of 4.6 million years is a very, very long time.

Despite the yearnings of humankind for information about extraterrestrial beings, there is almost nothing to support such an idea. Nevertheless there has been some thought-provoking research done on meteorites that opens up some interesting possibilities. Meteorites have been collected for centuries, but in the last few decades a proverbial "goldmine" of meteorites were located in Antarctica. Antarctica is almost entirely covered by a thick blanket of ice - a continental glacier. There are a few locations where the annual snowfall has not kept pace with the loss of ice by sublimation – solid water going directly into the gaseous state without being released as liquid water. Over the course of thousands of years, there are a few areas in Antarctica where the ice covering is totally absent. However, any meteorites that may have impinged on the old ice sheet are now exposed on the surface of the bare ground. Scientists have combed these areas and picked up thousands of meteorites. More meteorites have been discovered this way than in all the centuries before.

A thorough and exhaustive study of this treasure throve of meteorites was made. Most of them fell into the categories of meteorites that had been known before. However a small cluster of meteorites did not fit into these categories. Two different sets of meteorites were left over. One set was readily shown to be fragments from the surface of the moon. Astronauts had brought back numerous samples of moon rocks so this was easily established. However, the second set was different. Fortunately some of our space vehicles had landed on Mars and given us information about the composition of the Mars surface. A comparison of this set of meteorites with Mars composition indicated that we now have samples of the planet Mars. Meticulous examination of these precious samples showed that one Mars meteorite had tiny structures and compounds that inferred life may have been present on Mars.

This created a small sensation! Sober examination of these findings indicated that they could have been formed by inorganic chemical reactions, not necessarily by the life process. More Mars samples are needed to confirm or reject this claim for life. Plans for revisiting Mars are now being processed and hopefully additional information will shed more light on this interesting possibility.

Currently, there are planned expeditions by satellites to visit Mars and the moons of Jupiter, to try to discover if simple life formed on any of these bodies. If life did form there, that means that life can form readily. Life may be almost inevitable if conditions are suitable. In a sense, we are trying to determine if our speculations about intelligent life elsewhere have a firm basis in fact. We need facts to bolster or discourage any ideas we may have about life elsewhere. Satellite expeditions to remote regions of our solar system take time and a great deal of effort to accomplish. We have to wait and see what transpires.

Further suggested reading on biology and related topics

The Origin of Species, Charles Darwin, The New American Library, Inc., N.Y., 1958. (The original writings about the theory of evolution.)

Evolution, Frank H. T. Rhodes, Golden Press, N.Y., 1974. (A modern discussion of the topic.)

Darwinism Defended, Michael Ruse, Addison-Wesley Publishing Co., 1982. (An attempt to blunt criticisms of the theory of evolution.)

Darwin to DNA, Molecules to Humanity, G. Ledyard Stebbins, W. H. Freeman and Co., 1982. (The impact of modern science on the theory.)

Earth and Life Through Time, Steven M. Stanley, W. H. Freeman and Co., 1986. (The progress of Earth and life throughout the past.)

Life on Earth, David Attenborough, David Attenborough Productions, Ltd., 1979. (The history of all forms of life.)

The Living Planet, David Attenborough, Little, Brown and Co., 1984. (The story of Earth's surface and its colonization by plants and animals.)

The Trials of Life, David Attenborough, Little Brown and Co., 1990. (A natural history of animal behavior.)

The Blind Watchmaker, Richard Dawkins, W. W. Norton & Co., 1996. (A lively discussion in minute detail of the theory of evolution.)

Wonderful Life, Stephen Jay Gould, W. W. Norton & Co., 1989. (A detailed discussion of the Burgess shale fossils.)

Chapter 4

ANTHROPOLOGY AND ARCHEOLOGY

Biology is a broad field encompassing everything that lives. The focus is now going to be shifted to just one form of life - human. There are two scientific disciplines that study the past with respect to humankind - archeology and anthropology. They both intertwine to some extent and both provide insights into the development of humankind and our civilization.

Archeology is the scientific study of the life and culture of the past, especially ancient people, by excavation of ancient cities, relics, artifacts, etc. Anthropology is the study of the origin and physical, social, and cultural development and behavior of humans. Both of these disciplines are involved in the origin of humankind and how we got to be where we are now.

"Where did we come from?" is a thought-provoking question. Many people are so interested in this fundamental topic that they search back through records to find their immediate ancestors. A cousin of mine followed our family name back for about 250 years. But beyond that time, records were not available.

Bones and Artifacts

The same sort of thing happens when we look at humankind in general. We arrive at a point where the information is so meager that clear conclusions are not possible. In the recent past there are burials and written records to assist us. But as we go back in time, all written records disappear. But burials can still be found back to about 40,000 years ago. Few burials are ever found that are older than 40,000 years ago. As we go back even further in time, the information becomes more fragmentary. A few bones are found here and there. Some stone tools and campsites are located. Evidence for our ancestors becomes scattered and meager.

The earliest known human ancestors that have been located are about 4.2 million years old. These human-like creatures or hominids walked on two legs the same as we do. This is an important consideration, because gorillas, chimpanzees, orangutans, etc., have a grasping toe on their feet that is used for climbing trees. Humans do not have this feature. Further back in time, the fossil record is so incomplete that conclusions about which fossil belongs on the human ancestral line becomes mostly speculative.

An example is the ape-like creature, called Proconsul, that lived about 15 million years ago. Numerous bones of these creatures have been found, but there is no assurance that Proconsul was a human ancestor. Maybe he was and maybe he wasn't.

There have been two approaches to deciphering the development of humankind. One method is that of finding the bones, campsites, tools, etc., of primitive beings that were on the ancestral line. This, of course, is extremely important. Observing the bones, tools, etc., permits a detailed knowledge of how these creatures functioned. It provides some marvelous insights into how they lived and gradually developed into humankind.

Recently, an entirely new approach has been taken for understanding humankind and how we got here and when. It is a biochemical method. Careful analysis is made of various biochemicals within the human body. Proteins and DNA have been charted and minor differences noted. These differences and similarities make possible estimates of how closely related various human groups are to one another. It also allows a measure of how closely human beings are related to creatures like gorillas and chimpanzees and other animals as well.

A comprehensive examination of the biochemical method will be presented later in this chapter. Initially, the focus will be on the discovery of the bones and artifacts of ancient human-like creatures. There is a plethora of species names that tend to confuse and exasperate people who have not followed the discoveries closely. I have avoided the use of most of these terms and instead discuss the features that tend to make the specimens unique. Most of all, we must not lose sight of the fact that this is a continuum. The human ancestral line is an unbroken but ever changing sequence. There are side shoots that appear and then gradually disappear. But the human ancestral line never falters.

There are three distinct features to the human ancestral line that provide a common thread to follow. One is upright walking and the others are brain size and tooth structure. Human teeth are distinctly different from the teeth of gorillas or chimpanzees. One prominent feature is that both the gorilla and chimpanzee have large fangs, while modern humans have fangs about the same size as the rest of the teeth. Once upright walking began, a slow and steady increase in brain size took place. Eventually it culminated in the emergence of modern man, called Homo sapiens sapiens, who is we.

The entrance of modern man on the earth life scene was a very gradual development and took millions of years. The ancestors of modern man were not a dominant force when they first appeared on the life scene. But instead the improvement in the basic structure was gradual. Millions of years went by as modern man evolved from the primitive early forms. The main human ancestral line of evolution was but one process that occurred. Throughout this time period there were several lines of human-like evolution taking place. Human-like life forms appeared and then disappeared on the scene, making the fossil record a bit confusing. Sorting out the ancestral line has been very difficult.

Probably the best way to begin the search for the ancestral line of humankind is with the ancestor that led to the chimpanzee and Homo sapiens sapiens. This ancestral form, that lived about 6 million years ago, has not been located. There had to be ancestors before that, but the scant fossil hominid evidence is so fragmentary that any firm conclusions are not possible now. Today, the chimpanzee is the closest living relative to us.

There are two compelling bits of information that bring about this conclusion. The first is the marked similarity of the body structure and behavior of the chimpanzee when compared to human. Chimpanzees look, and in some ways act, a bit like we do. Then there is the biochemical information that shows the remarkable similarity of chimpanzee and human DNA as well as several key proteins. There is absolutely no doubt that the chimpanzee and human are closely related. The gorilla is definitely a more distant relative.

The earliest discovered fossil human ancestor lived about 4.2 million years ago. This creature walked upright and had some of the basic tooth structure of present day humans. About 3.5 million years ago this form had evolved somewhat into a slightly more human form. A moderate number of fossils of this type have been located, so that some compelling information about their walking behavior as well as their modification of bone structure indicated they were on the ancestral line of humans.

The finding of fossil footprints that are about 3.5 million years old was a most remarkable discovery. Careful examination of these hominid footprints showed that these hominids were walking with an upright, agile gait. These footprints provided solid evidence that upright walking was taking place long before modern man appeared on the scene

Unfortunately, several similar types of hominid forms lived during this same period of time, that were not on the human ancestral line. Even though these creatures walked upright, their tooth structure and brain dimensions did not conform to human-like properties.

The next human-like form on the ancestral line of humankind lived about 1.9 million years ago. The brain case was distinctly larger than any other form of hominid living during that time period. There is evidence that this hominid, called Homo habilis, was associated with stone tools. These were very primitive tools consisting of a hammer stone used to knock chunks from a flint like second stone. These sharp edged flint chunks were used as a cutting tool to get meat from a dead carcass. It is very difficult for a human being to penetrate the hide of a carcass using fingernails or teeth. But a sharp stone can slice through the hide, giving easy access to the meat. The sharp edged stones provided a new supply of food for these primitive people.

Bones, found in the immediate vicinity of these crude stone tools, show cut marks. These distinctive marks clearly indicate that the sharp stones were used to scrape meat from the bones.

About 2.4 million year old campsites of primitive man were located and numerous crude tools were found. Stone tools consisted of mostly choppers and scrapers. There is every possibility that Homo habilis used these tools, even though the bones of this human ancestor have not been found in the immediate vicinity.

The next ancestral form was a creature that had a body that was essentially a human body in shape and size. Only the brain case or skull was smaller than modern man. An almost complete skeleton, about 1.6 million years old, has been located. This form of hominid was a moderately sophisticated toolmaker. The tool kit consisted of flint objects that had a variety of uses. Hand axes, cleavers, and picks were made by this moderately large brained hominid. As time went by the brain size increased and so did the stone tool kit. Progress was slow but continuous.

Neandertal Man

About 300,000 years ago a new species of hominid appeared on the life scene, Homo neanderthalensis. Neandertal man was not the same as modern man. He was similar but distinctly different. Unquestionably, Neandertal man was present when evidence of modern man appeared in Europe about 50,000 years ago.

Neandertal man had an extensive tool kit. The variety of implements was remarkable. There were hand axes, projectile points, scrapers, knives, chisels, etc. The style of working stone was sophisticated and distinctive for Neandertal man.

At many sites, the bones as well as the tools of Neandertal man were found together. Neandertal man was the first human-like creature known to bury the dead. One specific burial involved considerable ritual: the dead man was carefully placed in his grave and flowers were buried with the deceased. It is an amazing discovery to learn that these ancient primitive beings acted very much as we do today. Significantly, the only creatures that bury their dead with ritual, are human beings, including Neandertal man.

Neandertal man was more stockily built than we are and possessed a slightly larger brain volume than modern man. How he disappeared is open to speculation, but we do know that Neandertal man abruptly became extinct. This dramatic change occurred about 35,000 years ago. Once modern man appeared on the scene, Neandertal man was supplanted or displaced in Europe in a very short period of time.

Neandertal man was the first ancient human-like fossil ever discovered. The discovery took place in Germany in 1856 and created a sensation. Ever since, there has been a great deal of speculation and discussion about the role of Neandertal man in modern man's development. Some people thought that Neandertal was definitely an ancestor, while many others did not think so. It was very difficult to determine the true nature of any possible relationship.

We know that Neandertal man lived in Europe, almost side by side, with modern man about 35,000 years ago. Yet about 30,000 years ago, Neandertal man had faded and disappeared. How and why this extinction took place is most unclear.

Recently, some biochemists chemically examined the bones of the first Neandertal man ever found. From these bones they were able to extract some mitochondrial DNA. Mitochondrial DNA is that special DNA that is found in every cell of the body and is involved in energy production. Mitochondrial DNA is only inherited from the mother. The father's mitochondrial DNA is not passed on to his offspring. The mitochondrial DNA stays constant for long periods of time, unlike the nuclear DNA which is scrambled at every mating.

Two different teams of researchers did the analysis of the Neandertal mitochondrial DNA. One was in the United States and the other in Germany. They obtained essentially the same results. Both groups were able to extract the same mitochondrial sequence composed of 328 nucleotides. In this Neandertal segment 27 differences were found from what is found in modern men of today.

Within living modern men from all over the world, there is only a maximum difference of about 8 positions in the same segment. Yet, Neandertal man has 27 differences. This is a much larger number of differences. The mitochondrial DNA of modern man and Neandertal man are distinctly different. The conclusion is that Neandertal man is a most unlikely candidate as an ancestor of modern man.

Unfortunately, the work was done on just one specimen of Neandertal man. To be more certain about the conclusion, additional analysis is needed of other Neandertal specimens. If the additional data gives similar results to those already obtained, then the conclusion that Neandertal man was not an ancestor would be confirmed.

This information now indicates that Neandertal man was not an ancestor of ours. Additional biochemical information indicates that Neandertal man and modern man had a common ancestor about 600,000 years ago. Modern man first appeared in Africa about 200,000 years ago. All indications are that Neandertal contributed little, if anything, to the genetic inheritance of modern man.

Numerous efforts have been made to try to decipher what caused the demise of Neandertal man, but nothing definitive has been discovered. There has been a lot of speculation about what took place, but there are no findings that support any of them. The most plausible idea seems to be that modern man beat out Neandertal man in competition for the same resources. There is also the unsubstantiated idea that modern man made war on Neandertal man. Around the world for several million years, the ancestors of modern man persisted while competitors disappeared from the scene. There is the unpleasant conclusion that the more sophisticated hominids destroyed the less sophisticated. But so far no firm evidence of genocide has ever been found.

Modern Man

Once modern man appeared, the stone tool kit expanded considerably in a relatively short period of time. Startling new activities occurred. Painting, sculpturing, and carving, totally unlike anything seen before, became part of human culture.

New endeavors became almost common. Spears and arrows became routine weapons. Animals were domesticated, agriculture began, dwellings were built and villages developed. Boats that were capable of traversing fairly large distances at sea were invented. Modern civilization had started.

Then metals were discovered. copper and bronze were the first metals used extensively. Tools and weapons were made of metal rather than stone. The metal provided a much sharper cutting edge than stone. Soon iron replaced these softer metals and modern civilization moved another step closer to developing our world of today.

Iron was very versatile and made weapons that were much more durable and retained a sharp edge. Iron also found uses in buildings and household equipment, like pots and pans. Man left the stone age and entered the iron age. In a sense, we are still in the iron age today.

Progress toward a civilized world continued, even though warfare continued to break out between various factions. Such strife is still all too common today. However, the intellectual activities of humankind continued to flourish, and significant changes continued to be made.

The last thousand years has been an almost frantic pace in the way life has changed. Numerous advances in engineering, navigation, construction of elaborate buildings and bridges, have taken place. The intellectual surge in human activities has been remarkable. Music and the theatre became firmly established. Sculpture and painting reached new heights.

Two important inventions occurred that changed humanity. Both were a new form of communication. The importance of these can be easily overlooked. But the development of writing and then of the printing press were of profound importance. Up to the development of writing the only important means of communication were speaking and gesturing. These were spur of the moment processes that had permanence only in the brains of others who heard or saw the communication.

But writing became a new and semi-permanent means of communication. The thoughts and wishes of an individual were no longer just transitory, but became a semi-permanent record of what had transpired. The thoughts and ideas of an individual could be communicated even after that individual had died. Facts and information could be communicated in a semi-permanent manner. This increased humankind's store of knowledge immensely.

The development of the printing press made the record of information much more permanent than it had been. Instead of one copy of written information, thousands, even millions could be made. Information could be distributed over long distances and to many individuals. Written information is still semi-permanent, but it much more secure than it was before printing became commonplace.

Right now we are in an almost unbelievable surge of technology. We have trains and planes, automobiles and ocean liners, radio and television, telephones and computers, synthetic fibers and plastics. Electricity has been developed as well as nuclear power and atomic weapons. Recently, humankind has moved at an astonishing rate into the high technology of today.

Human beings have set foot on the moon and our satellites have visited many of the interesting objects in our solar system. We have various telescopes in orbit above the Earth's atmosphere sending us fine details about stars, galaxies, and the depths of space itself. Here on Earth, huge telescopes are gathering data on numerous objects in outer space. We now have information about neutron stars, black holes, colliding galaxies, etc. The pace of discovery is astounding.

We tend to forget the simple beginnings and the painstaking slowness of the advance of technology. But developing our complex society actually took millions of years. A million years ago the change in technology was worse than a snail's pace. In periods of hundreds of thousands of years perhaps a few stone implements might have been added to the tool kit of humankind.

Modern man changed all that dramatically. In the course of about 50,000 years humankind evolved from a simple hunter-gatherer mode of life to our modern day civilization. The information acquired in the last few centuries is enormous.

We found out that the Earth is spherical and spins on its axis, and that the Earth revolves around the Sun. New worlds, like North and South America, Australia, and Antarctica have been discovered. Detailed maps of the Earth's surface as well as the sea floor have been obtained. Progress has been very substantial. Today, humankind has a good idea of just how we fit into the overall scheme of things.

More Genetic Information About the Origin of Man

Until about 30 years ago, information about the development of humankind was almost exclusively in the realm of archeology and anthropology. But significant advances in biochemistry introduced some new methods for deducing how humankind has developed. Studies of the biochemistry of living organisms showed the similarities as well as the differences that were present.

Some original work with proteins and human ancestry provided the opening wedge to an entirely new way of looking at humankind's development. This work was then expanded into studies involving DNA. Some new and very clear insights were presented by this innovative approach. Creatures living today, not fossils, were the major source of the specimens studied. Changes in the makeup of their biochemicals provided definitive information.

One extremely informative biochemical is DNA - deoxyribonucleic acid. It is a component of every cell of our bodies and is the basis for genetics. DNA determines the inherited characteristics of every individual. There are two types of DNA. One is nuclear DNA - a very large and complex molecule of about 3.2 billion base pairs. There are about 30,000 genes in nuclear DNA. Genes determine our hair color, our height, our muscle structure, all our features, etc. The second form of DNA is mitochondrial DNA - a relatively small molecule when compared to nuclear DNA. Mitochondrial DNA is a component of the mitochondria - minute globules found throughout the cells. The mitochondria generate the energy needed by the cells. Within the mitochondria are found a small DNA that has but 37 genes. This is much more manageable to study than the 30,000 genes of the nuclear DNA.

Nuclear DNA is inherited from both parents, but mitochondrial DNA is inherited only from the mother. There is no complex shuffling of genes as there is in nuclear DNA. The mother makes the only contribution. This makes following ancestry much easier.

DNA undergoes mutations. Over the course of time, the composition of the DNA molecules becomes changed by the action of cosmic radiation, natural radioactivity, and chemical mutagens. These changes are random, but become statistically predictable over long periods of time, such as thousands and millions of years. So, the longer a DNA molecule has been in existence, the more changes will occur in the chemical structure, the base sequencing, of the DNA. This is the basis for the DNA clock.

The mitochondrial DNA has relatively short strands of base units when compared to nuclear DNA. The study of chemical changes in mitochondrial samples from various organisms makes a reasonable method for determining elapsed time. Few changes indicate a short time period. Many changes indicate a longer time period. Mitochondrial studies indicate that about five to seven million years ago, the chimpanzee and humans had a common ancestor. When biochemists compare human and chimpanzee mitochondrial DNA they find that there have been numerous changes in the chimpanzee mitochondrial DNA as compared to human mitochondrial DNA. These changes occurred since the chimpanzee and the human had a common ancestor, about five to seven million years ago.

From the differences between the mitochondrial DNA of the chimpanzee and humans, and the five to seven million years it took for these changes to take place, a rate of change was established. Over the course of a million-year period, the rate of change in mitochondrial DNA was found to be from two to four percent. This has a fairly large uncertainty range.

When human mitochondrial DNA is studied, there is only a maximum change of 0.57% between the many human beings examined. Individuals from all the continents were studied. No large ethnic groups were left out. If human DNA had been around a million years we would expect to see a two to four percent change. So, human beings have diverged from one another for much less than a million years. The mitochondrial DNA of modern humans had to originate about 200,000 years ago. The uncertainty is about plus or minus 100,000 years. Once again, a fairly large uncertainty range is present.

This date gives us some idea of when modern man first emerged. The mitochondrial DNA of every human being in the world today is all very similar. It makes no difference whether your immediate ancestors came from Europe, Asia, Africa, Australia, or the Americas. The mitochondrial DNA differs within a maximum of 0.57%. All human beings are brothers and sisters under the skin. Our mitochondrial DNA all came from the same woman (or a small group of closely related women) about 200,000 years ago. She was, or they were, the source of the entire human race's mitochondrial DNA - a most interesting and thought-provoking conclusion.

Even though one female, or a small group of females, were the source of our mitochondrial DNA, not all our characteristics can be traced back to them. But our mitochondrial DNA certainly can.

An analogy, with family names, can help to understand how this came about. In our society, the children inherit the family name of the father. The family name of the mother is not retained. If the assumption is made that every married couple has but two offspring that eventually mate and have families, then the analogy will demonstrate how humankind ended up with the mitochondrial DNA of but one ancestor.

Start with eight couples. The family names of the eight women are not retained. The family names of the males have the potential of being retained if male offspring result. Statistically, after the first series of offspring only six family names will be left. That is because two of the eight couples would have only girls.

I continued this procedure with coin tossing to determine the sex of the resulting matings. I ended up with but three family names from the original eight family names after five generations.

Another set of coin tossings might give a slightly different result. Nevertheless, family names tend to be eliminated over long periods of time.

The same sort of thing happens with mitochondrial DNA. Let's begin with eight couples and assume that each member has entirely different mitochondrial DNA. The eight male mitochondrial DNAs are lost at the first mating. Only the female mitochondrial DNA is retained. Once again we will assume that only two offspring from each mated pair reach adulthood and can have offspring. At the first mating two female mitochondrial DNAs are lost because statistically two couples will have only boys. As the matings continue there is always a chance that some mitochondrial DNA will be lost. A period of 200,000 years represents about 10,000 generations. That is a very large number of chances for elimination of mitochondrial DNA. The probabilities suggest that sooner or later only one mitochondrial DNA will persist.

A study has been made of the DNA in the Y chromosome of men. This study took eight years to complete. This study was much more difficult than the mitochondrial DNA study. The DNA sequences of the Y chromosome are much larger than the entire mitochondrial DNA sequence. In order to simplify the study, only one gene of the Y chromosome, the ZFY gene, was studied. This gene is inherited from father to son, and does not undergo any scrambling when the male and female chromosomes merge.

The base unit sequencing of the ZFY gene was determined for various men around the world as well as for the gorilla, orangutan, and chimpanzee. The analysis of all the data collected indicated that the Y chromosome of modern man was inherited from either a single male or a small population of males that lived about 270,000 years ago. The ZFY gene of the Y chromosome of modern man had descended from a recent, common ancestor.

The two studies, of the mitochondrial DNA and the ZFY gene, reached similar conclusions. Both studies indicate that the ancestors of modern humans originated about one-quarter of a million years ago. Even though there is some statistical uncertainty in the dating, the two studies are in general agreement. Modern human beings have a recent and common ancestry.

There has been an hypothesis that the modern "races" of human beings all developed independently of each other, by modification of the Homo erectus form. This new information indicates otherwise. Even though Homo erectus had scattered around the world, modern man, Homo sapiens sapiens, eventually replaced Homo erectus, in all locations.

Unquestionably, modern man did develop from Homo erectus, but only once in Africa. Then modern man scattered around the world and replaced the less sophisticated forms of hominids, including all Homo erectus in all parts of the world.

Furthermore, every human being has almost identical DNA sequencing in his or her mitochondrial DNA. A similar statement can be made for every male human being about the DNA sequencing of the ZFY gene on his Y chromosome. It is almost identical. All of humankind is closely related. We are all brothers and sisters under the skin.

A great deal of work has been done on the analysis of proteins within the human body. One of the earliest studies involved the beta chain of hemoglobin. All proteins in the life system of Earth make use of but 20 amino acids in their construction. The beta chain of hemoglobin contains 146 amino acids in a specific sequence. Even though one of each of the 20 amino acids can appear in each of the 146 positions, the beta chain of humans has a specific sequence. Each position contains a definite amino acid. All 146 positions are identical in all healthy human beings.

When a comparison of the sequence of amino acids in the beta chain of hemoglobin was made between the human and the gorilla, only one difference was found. The biochemistry of the gorilla and the human are very similar.

However, when the comparison was made between the sequence in the beta chain of hemoglobin for the common chimpanzee and humans, the two sequences were identical. This indicates that humans and the chimpanzee are very closely related. Even though a human is related to the gorilla, the relationship to the chimpanzee is closer. Studies of other common biochemicals show similar striking resemblances between the chimpanzee and humans. The biochemical information continues to indicate that the chimpanzee and humans are closely related.

Studies of proteins, like the beta chain of hemoglobin, between various animals show the relationships between them. Such analysis shows how a sheep is related to a goat, or a horse to a zebra, etc. Numerous differences indicate a fairly distant relative. Few differences indicate a close relationship. There is no need to dig up fossils or to date rock sequences. The creatures are alive now. The relationship of humans to a chimpanzee can be determined over and over again. Any healthy human being is suitable for the study, including you.

Very recent studies of the DNA sequencing between various different organisms has disclosed some definitive results. For example: the relationship between the chimpanzee and humankind can now be mathematically stated. The DNA of the chimpanzee and of humankind is 98.5% identical. Out of 1000 positions in the DNA chains, only about 15 are different. The DNA work in progress is decisively demonstrating that the theory of evolution is correct. This work clearly shows the relationships between all kinds of organisms. And most importantly, relationships can be mathematically expressed. Guessing has been almost eliminated.

The theory of evolution predicts that the more dissimilar two organisms are from one another, the more differences there should be in their DNA. That is what is found when the DNA of various organisms is analyzed. Human DNA has more differences with respect to dog DNA than human DNA does to chimpanzee DNA. Bird DNA has more differences than does dog DNA to human DNA, and fish DNA has considerable more differences when compared to human DNA than bird DNA does. When human DNA and worm DNA is analyzed, the differences are the highest of the five species studied, just as would be expected from the theory of evolution. Observation of this type confirms the validity of the theory of evolution. The longer the time period since two species diverged, the more differences that appear in their respective DNA.

If we examine the creatures mentioned, human, chimpanzee, dog, bird, fish, and worm, we would probably arrange them in the order given here. After all a dog is considerably closer in anatomy to a human than to a worm. The DNA just confirms what we would deduce from the anatomy considerations. It also gives us a mathematical value for comparison. So the DNA studies have reaffirmed the fundamental ideas in the theory of evolution. The idea that evolution has happened is compellingly reinforced.

One other aspect of the DNA work in progress is the gene manipulation studies. Various organisms can manipulate human genes and produce human molecules. That implies a very close relationship in the biochemistry of all of them. Biochemists have successfully inserted human genes into various other organisms, such as: bacteria, mice, goats, and cows. These widely diverse creatures have been able to produce human substances that can be used in medicine. A specific example is the insertion of the human gene for producing human insulin into bacteria. The bacteria are able to turn on the human gene and produce human insulin. Large batches of bacteria can be grown so that the human insulin can then be harvested and used to treat individual humans who suffer from diabetes.

Prior to this DNA work the treatment for diabetes was the introduction of either beef or pork insulin into the diabetic patient. This beef or pork insulin was obtained from the carcasses of cattle or hogs in slaughter houses. Some diabetic patients could not readily tolerate this animal insulin. The supply of insulin was limited by the total number of animals that where slaughtered. Now there is an unlimited supply of human insulin obtained at a lower cost. This represents a profound improvement in the treatment of diabetes.

In the gene manipulation studies, many urgently needed human substances are being produced by various creatures. These compounds will be used in medical treatment. Insulin is only one of the many badly needed human substances that have been made. In some cases the desired material is secreted in the milk of cows or goats and the needed substance is extracted from the milk for medical use. The term "pharm animals," as in pharmaceutical, is used to describe these highly specialized farm animals. The possibility of expanding this approach looks very promising.

But the most important findings from all this work is that all people on the Earth are very closely related to one another. The DNA results indicate that all humans everywhere have DNA that is over 99.9% identical. There can be no doubt about it. Every human being is very closely related to every other human being on the Earth. The idea of racial differences is due to superficial features such as the color of the hair or the color of the eyes. From a scientific point of view there is only one race of man - the human race.

The classification of various ethnic groups as different races is just plain wrong. There is no such thing as a black race or a white race. There are human beings that have slightly different pigmentation of their skin but that most certainly does not make them a different race. The major differences between people are cultural. Some speak different languages, have different religions, and wear different types of clothing, but that does not make them a different race. They have learned to behave differently. True racial differences are not present in their DNA. All the so-called races of man can interbreed without any difficulties. Which decisively demonstrates that all humankind are closely related and not separate races.

If someone mentions anything about the races of man, you will instantly know that they have not kept pace with the latest scientific findings in biochemistry, which makes the concept of races of mankind obsolete. From a biological point of view there is only one race - the human race. This concept needs to penetrate our politics, religions, social attitudes, and most of all our daily viewpoint of each other. Calling someone a racist becomes blurred by the knowledge that there is only one race. Hopefully this one scientific fact - that there is only the human race - will eventually sink into the consciousness of everyone in the world and fundamentally change our way of looking at one another.

Further suggested reading on archeology and anthropology

The Making of Mankind, Richard E. Leakey, E. P. Dutton, N.Y., 1981. (A survey of the fossil evidence.)

Lucy, The Beginnings of Humankind, Donald Johanson & Maitland Edey, Warner Books, N.Y., 1982. (A detailed discussion of field work in searching for human-like fossils.)

The Search for Eve, Michael H. Brown, Harper & Row, N.Y., 1990. (Biochemical information demonstrating the origin of humankind.)

The Origins of Humankind, Richard Leakey, Basic Books, N.Y., 1994. (A survey of human development.)

African Exodus, Christopher Stringer and Robin McKie, Henry Holt, N.Y., 1996. (The development of modern man and his journey throughout the world.)

From Lucy to Language, Donald Johanson & Blake Edgar, Simon & Schuster, N.Y., 1996. (Marvelous pictures of human-like fossils with discussions about them.)

Origins Reconsidered, Richard Leakey, Doubleday, N.Y., 1992. (Explores how we became human.)

The Neandertal Enigma, James Shreeve, William Morrow and company, Inc., USA, 1995. (A thorough discussion of Neandertal information plus analysis of many other human-like fossils.)

The First Humans, American Museum of Natural History, Weldon Owen, Harper Collins, N.Y., 1993. (Human Origins and History to 10,000 BC.)

People of the Stone Age, American Museum of Natural History, Weldon Owen, Harper Collins, N.Y., 1993. (Hunter-Gatherers and Early Farmers.)

Chapter 5

THE PRESENT

The world of today presents some obvious difficulties that require concerted action on a worldwide front. Involved in these difficulties are long established social and religious customs and attitudes that need to be courageously confronted and changed. Many of these old out-moded practices must be abruptly stopped or they will continue to plague mankind's future on this planet.

The biggest worldwide problem is the expanding human population. We are now over 6 billion human beings in the world, and there is every indication that the human population will continue to increase into the indefinite future. Right now we are adding about one million persons to the world population every four days. This increase has been going on for several decades and shows no indication of changing dramatically. The demands made on the environment by the new inhabitants are troublesome. Every four days these new one million human beings need additional food, water, space, etc. The pressure on the environment is relentless. Eventually this population pressure will become too great. Eventually something has got to give. We can anticipate that famine, disease, warfare, revolution, anarchy, etc., might appear in response to the pressure of too many people.

The obvious change in the world's climate is another problem that demands attention. Before pre-industrial development, the atmosphere did exhibit a greenhouse effect. Water, carbon dioxide, methane, nitrogen oxides, and a few other gases absorbed heat radiation to warm the atmosphere. However, our present industrial civilization produces considerably more of these greenhouse gases and in response the atmosphere of the earth is slowly getting warmer. Not only the land is involved but the oceans as well.

The oceans occupy slightly over 70% of the Earth's surface. As the atmosphere experiences a rise in temperature, the contact between the atmosphere and the hydrosphere ensure that the temperature of the oceans will correspondingly increase. Higher ocean temperatures will drastically affect the ecology of the marine world.

There are strong indications that the enhanced greenhouse effect is already operating significantly. The pattern of rainfall worldwide has been altered in erratic ways. There are droughts in areas where such conditions were almost non-existent. Tropical storms have been more numerous and more vigorous as well. Surface temperatures are increasing. Humankind can do something about this, but drastic worldwide action must be taken to control the situation. So far control of worldwide pollution has not made any significant progress.

World health is almost at new heights, despite the onslaught of AIDS. There are still persistent deaths from malaria and similar tropical diseases. But the overall health picture for humanity is encouraging. Smallpox has been eliminated.

One significant indicator of world health is that the life span of individuals has significantly increased. Vaccinations and anti-biotics have contributed in a major way to overall worldwide health.

The wide spread use of vitamin pills and similar medication has enhanced life. The daily taking of half an aspirin tablet has been shown to prevent about 50% of strokes and heart attacks. Ingesting a daily vitamin E pill appears to prolong life. The benefits of such simple steps have contributed to the increased life span.

Warfare is still in vogue, although the overall toll is considerably lessened from what it was just a few decades ago. World War III has not taken place - a most encouraging development. My three sons and three daughters did not have to fight in World War III. This is of overwhelming importance to me. Perhaps humans are learning other ways to settle old differences.

The type of warfare that is rampant is often of a sectional nature. Civil wars keep breaking out among groups of people that have a long history of strife. Border squabbles still take place, and revolutions routinely occur. There are also internal tensions where dissenting groups attempt to unseat existing governments. Armed violence is all too common under these conditions.

The status of women throughout the world still has a long way to go before reaching equality with men. Throughout human history, men have dominated politics, religion, warfare, commerce, etc. Only in fairly recent times has it been possible for women to move occasionally into positions of power or authority. In the industrialized countries women are now approaching the status of men. They can vote in elections, own property and seek public office. Progress is continuing, but the situation is still far from satisfactory. In the United States no woman has ever held the office of president or vice-president. Women have rarely become governors, and only about 10% of the United States national legislature is female. This is very unsatisfactory because women represent over 50% of the eligible voters in this country. In a democracy, the governing body of the nation should be representative of the population governed and right now that is not true. Part of the democratic process is being thwarted.

However, in the developing countries, the status of women is about the same as it was in the dark ages. The importance of the status of women cannot be overemphasized. In countries where the status of women is relatively high, the birth rate is low. In countries where the status of women is low, the birth rate is high. The correlation between the status of women and birth rate suggests that this is one way to bring about a reduction in human population. This problem of over-population must be faced, and soon. Increasing the status of women may provide a means of accomplishing lower birth rates. It certainly will not make things worse.

Even though a great many procedures for lowering the birth rate have been proposed, the most direct way to lower the population explosion is to increase the status of women worldwide. This means that social and religious customs around the world must be drastically altered. Women are now forced to wear a veil in some countries. This custom has got to go. A lot of other restrictions on women's behavior must also be eliminated. Women must be allowed to own property and be free to move about as they please. Women must be given the right to vote, to drive cars, and do everything that men are permitted to do.

In some nations men oppose giving women any rights that will improve their present status. Men tend to resist any changes that threaten the male's present position of domination. Even a somewhat modern appearing country, like Saudi Arabia, imposes dogmatic authority to continue the subjugation of women. Just before the Gulf War, American soldiers were driving trucks around Saudi Arabia in order to distribute needed military supplies. Some of these truck drivers were women in the United States military forces. This presented a problem, because in Saudi Arabia women have never been allowed to drive motor vehicles. Some western educated Saudi women decided this was an opportunity to reverse this age-old tradition. These women had driven motor vehicles in other countries where they had studied. To protest against the edict that women could not drive, a group of Saudi women drove cars about Saudi Arabian government buildings. These women were immediately arrested and discharged from their places of employment. These educated women included teachers, lawyers, doctors, and scientists who were punished without trial for violating the edict that women

could not drive. They were denounced for their "intolerable act" -
driving a motor vehicle.

Incidents of this type are all too common in some countries.
Entrenched ideas are not dismissed easily. The response of the
dominant males indicates that liberating women worldwide will
not be easy to accomplish. Nevertheless, the pressure on the
world's resources from increasing world human populations does
not give time for bringing about drastic change slowly. There
needs to be an abrupt improvement in the status of women, and
the sooner the better for the entire world.

Once women on a worldwide basis are given their basic rights,
the birth rates should become rational. Reproduction worldwide
would then approach what it now is in the developed countries.
Currently, the undeveloped countries are just about the only major
contributor to excess world population.

The pressure of increasing world human populations is driving
many species of plants and animals around the world to the brink
of extinction. The only sure way to preserve endangered species is
to limit the human population. A lower human population would
also produce less pollution. That would also benefit wildlife. The
most direct way to achieve these objectives is to increase the
status of women. The solution seems so simple, but it comes
smack up against entrenched social and religious customs. The
culture in many countries is based on the subjugation of women.
Religious restrictions are a major part of the problem.

Religious differences continue to be a bane on humankind. So
many of these ancient ideas are divisive. They must be rethought.
If humankind is ever to live in peace and tranquility, the problems
of religious differences must be overcome. As it is now, religious
attitudes seem to provide a basis for confrontation that leads to all
sorts of vicious strife.

The number of various religions is extremely high. The world
religions are in a state of chaos because of so many different ways
of doing things. If one ponders the situation some obvious
conclusions are possible. All of these religions annot be on the

right track. Perhaps one of them might be, but all of them cannot be right. There is also the distinct possibility that none of them are correct.

It is also obvious that many of these religious ideas should be eliminated as unwise. Tradition and social customs are often offshoots from religion. Examples are: the wearing of veils by women, and the so-called female circumcision (also called genital mutilation). Both of these customs subjugate women to a lesser social status. Female circumcision is positively barbaric and is a serious violation of basic human rights.

Religions in their present form cannot be as sacred as they all claim. The multitude of beliefs, procedures and customs are all too often out-moded, divisive, and unsupported by rigorous facts. Religions tend to divide humankind in an artificial way. Almost all religious factions are without any firm scientific basis for supporting many of their religious ideas. Much of religious dogma falls into the realm of non-science and occasionally falls into the category of nonsense. Examples are the taboos about eating meat. Some consider it a sin to eat any type of meat on Fridays. Some forbid the eating of pork anytime. Such rules are without any valid modern day justification.

Chapter 6

THE FUTURE

Astronomy

When astronomy is examined carefully, the future of the Earth appears to be very secure. Unfortunately some purveyors of doom keep predicting that the world will end soon. Numerous religious groups have proposed earth destruction, over and over again. But the Earth does not end. On the designated date of destruction, the Earth moves about without any hint of a problem. From a scientific point of view, the total demise of the world appears to be extremely unlikely in the near future and even in the moderately long future.

The procession of various religious figures predicting the demise of the world has brought profound fear and alarm in believers. Many individuals have taken frantic steps to prepare for the worst as predicted by their trusted religious leaders. But when the date of supposed destruction arrives, the Earth continues on its normal way, totally oblivious to the dire prediction. So when the next prophet of doom gives a date for Earth destruction, ignore it. There is no reason to believe that the end is imminent.

Astronomers have examined the immediate surroundings of the solar system in outer space and have found nothing sinister. There are no nearby stars ready to undergo a supernova explosion. There are no nearby stars on a collision course with the Earth.

There are no calamities for total Earth destruction awaiting us. And we must remember that the Earth is already 4.6 billion years old. The Earth has been, and still is, in a stable part of the Milky Way galaxy. There are no indications that "the end" is in sight.

Nevertheless, the inevitable end of the solar system will occur. In about 5 billion years, the hydrogen fuel that powers our Sun will become depleted. Then our Sun will undergo a dramatic transformation into the red giant phase. The Sun will balloon outward and engulf the planets of Mercury and Venus. The Earth will probably escape being engulfed, but it will be thoroughly roasted. The oceans will evaporate, the surface temperature will climb, and all life will perish. Our Sun is doomed to die and in its death throes it will destroy life on Earth. There is no way to escape this final disaster. But remember that is about 5 billion years into the fixture. Humankind has only been on the Earth for about 50,000 years. The eventual destruction of the Earth is in the very distant future.

There are still some problems to surmount with respect to preserving humankind. Those involve possible collisions of the Earth with large asteroids or comets. Such encounters have occurred with moderate frequency in the past. The last known large collision occurred about 65 million years ago. This is believed to have caused the extinction of all the dinosaurs and many other forms of life as well. Such large impact encounters are almost certain to happen again. The Earth does not get destroyed in such a large impact, but life forms are obliterated by the impact as well as the aftermath of such a huge collision. In the past, some life forms have always survived. Yet there is a distinct possibility that such a future impact could exterminate humankind.

The aftermath of such an impact could be horrendous. So much debris would be introduced into the upper atmosphere that most sunlight could not penetrate to the Earth's surface. Plants could not use the photosynthesis reaction that requires sunlight to power the reaction. Plants would die and the animal life that feeds on plants would have little to eat. The entire food chain would be totally disrupted. A "nuclear winter" type of chilling would take place all over the world. Surface temperatures would become much like winter, even in the tropics. This would have a devastating effect on both plant and animal life. Humankind is very resourceful, but there can be no guarantee that humans will survive a huge impact event and its after effects.

However, humankind is not entirely helpless with respect to possible large extraterrestrial collisions. Even though such collisions have occurred several times in the last 500 million years, they can be circumvented. We have already surveyed the confines of the solar system and have found most of the really big asteroids. We know their orbits and none of these are candidates for a collision in the moderately distant future. However, we might have missed a few, and any one of those missed could cause a major calamity. But that is not very probable.

Nevertheless, there is the distinct possibility that a comet could collide with the Earth. Some of these comets are big enough to be a profound threat to our world. These originate in the far reaches of the solar system, beyond the orbits of the planets. It could be possible that a comet, totally unknown to astronomers, could enter the inner solar system and be on a collision course with the Earth. This would be a worrisome challenge for humankind. With our space capability and extremely powerful hydrogen bombs we should be able to do something that will limit the destruction. Such a comet-Earth collision is a very remote possibility. Nevertheless, scientists have considered such a scenario and have made some tentative plans for thwarting such a collision.

If there is a fairly long period of time, before the collision would occur, a space mission could land on the comet and attach a nuclear engine that will slowly push the comet into a path that will miss colliding with Earth. But if the time is short, a powerful hydrogen bomb could blast the comet into small fragments that will become spattered all over the volume of space about the comet. Many of these will fail to collide with Earth, although plenty of them will collide. The collision will not be one impacting event, but trillions of small fragments and individual molecules entering the Earth's atmosphere. The destruction would be significantly reduced. Perhaps human life might survive such a dissipated impact.

In terms of the survival of the entire Earth, there is not much else to worry about. We must remember that the solar system has been operating for 4.6 billion years without any catastrophe. There is no known reason why it cannot continue to operate for another 5 billion years.

But before leaving the astronomy, I want to mention one bothersome possibility, a collision with a small asteroid or comet. The Tunguska event in Siberia in the early 1900's was such an encounter. Humankind should try to locate all the significantly sized objects that might impinge on our planet and cause a Tunguska-like event. The Tunguska event occurred in an uninhabited area of Siberia and no human beings perished. But if the event had occurred in a densely populated area, many thousands of human beings would have been killed. So some careful scrutiny of all these relatively small objects would be a prudent course to follow.

The meteor crater near Flagstaff, Arizona is another example of a modest strike by an extraterrestrial object. This was caused by an iron meteorite and gouged out a crater that is almost a mile across. This gives some indication of how devastating such a strike might be if it occurs in a populated area. Thousands of individuals would die. The devastation would be similar to an atomic bomb attack.

Geology

The future of world geology also presents some worrisome problems. By far the biggest is the eventual melting of the icecaps of Antarctica and Greenland, as the world warms from the enhanced greenhouse effect. About 2.15% of the world's total supply of water is now present, as ice, on land. This solid water, when melted into liquid water, will raise ocean levels considerably. Added to this will be the thermal expansion of the oceans as the world's atmosphere gets warmer and raises the temperature of the seas. The total rise in sea level has been estimated at between 200 and 250 feet. This will take a long time, perhaps many hundreds of years to bring about all this melting. But the ultimate end is that all the ice will eventually melt, unless humankind takes some sort of drastic action to limit our pollution of the atmosphere. Currently, humankind is adding over 23 billion tons of carbon dioxide to the atmosphere every year. We are also adding other greenhouse gases such as: methane, chlorofluorocarbons, nitrogen oxides, etc., to the atmosphere. All these gases contribute to the enhanced greenhouse effect that is bringing about global warming.

A rise of about 200 to 250 feet in sea level will inundate all the major seaports throughout the world, as well as all the low lying coastal areas. Bustling cities like London, New York, Singapore, Tokyo, Rio de Janeiro, Cairo, Amsterdam, etc., will vanish beneath the waves. The economic and social upheaval caused by such a series of debacles is difficult to imagine, but it will be almost earth shaking. And at the same time much of the good agricultural land that helps to feed the human population will also be taken out of production by the rising waters. There can be no doubt that the impact of rising ocean levels will have a profound effect on humankind all over the world.

The change in worldwide geology because of the rising ocean levels will change national boundaries drastically. For example: The Netherlands and Bangladesh will disappear from the map. The state of Florida in the United States will also disappear. These present day bodies of land will become ocean bottom.

Biology

The enhanced greenhouse effect will strike in a variety of ways. Rising oceans levels are but one of the many ways global warming will change our world. As the world continues to get warmer, tropical diseases will slowly and relentlessly invade the temperate regions of the world. Malaria kills about 3 million human beings per year right now. That figure is guaranteed to rise as the world's weather get warmer. There are numerous tropical diseases, and human beings in the temperate regions of the world are not prepared to cope with them. The toll will be extensive.

Tropical plants will also invade the temperate regions of the world. The Earth's biology will undergo a drastic alteration as temperate plants are supplanted by tropical varieties. All forms of life in these regions will be under stress. Not only will trees, bushes, and vines have a difficult time but so will grasses, soil micro-organisms, insects, amphibians, reptiles, etc. Unquestionably many life forms will become extinct. The upcoming changes are taking place too fast for organisms to respond and adapt in order to survive.

Biology throughout the world is in for a very bad time. Humankind is taking over almost the world's entire surface, there is little habitat left for numerous species of plants and animals. As the human population continues to expand, the area available for all other forms of life shrinks. Added to this human population pressure is the stress from climate change. The outcome is obvious. Extinction will become all too common. The world's roster of living organisms will continue to decrease. From a biology point of view, the world will become a much poorer place. In the very near future, human beings will have taken over practically all the surface of the globe to meet their needs, leaving little if anything for the other forms of life. There will be scant space left for wild creatures.

Archeology and Anthropology

What will happen to humankind? There will be so many people on the Earth that the food and water supplies will be insufficient to sustain them. There will be the formidable stress of too many people and not enough resources to go around. Of course we can circumvent this debacle by limiting our world population growth now, but there does not seem to be any concerted worldwide effort to restrict births. Unless a sustained worldwide system to control the population is implemented, the specter of too many people will plague humankind.

Humankind has conquered many diseases. Now individual humans live longer than they did before. Considerably more people are living on the Earth today because of improved human health. However, our biological reproductive urges are still as virulent as they have been down through the centuries. We now have some excellent birth control procedures, but the vast majority of humanity cannot afford them or are ignorant of them. As a whole, humankind does not have the will or the means for controlling reproduction worldwide. So I would guess that we will continue down the path to drastic over-population until hunger or disease brings us to our senses.

Humankind has become the supreme masters of the globe, but unable to control themselves. Humankind should be able to control the destiny of the world, but humankind does not seem able to accomplish it. World wide sustained action has not been a strong point for human beings. Perhaps we might do better in the future, when we recognize the situation more clearly than we do now.

Let's finish on a promising note. The unraveling of the human DNA sequences places us in a position to control our personal destinies. Thousands of human genetic diseases could be eliminated by careful selection of healthy genes or unwholesome genes. This will require some losses, and there will be cries of foul from religious leaders and social groups. The long-term benefits to the human race outweigh all the breast beating and shouting. Disease carrying genes can be eliminated.

I remember a situation that developed in a cattle herd owned by my father in law, Walter S. Gardner. Dwarf cattle began to be born. Dwarfs presented both a financial problem and a handling problem for a rancher. In order to eliminate these problems my father-in-law sold every bull and every cow that had anything to do with the dwarfism. Within two years his herd no longer had a dwarf problem. He had removed the "bad" genes responsible for the problem.

Eliminating "bad" genes on a human scale is considerably more difficult. We do not have the option of permanently removing possible participants. Currently, one option is now available - early abortion could eliminate undesirable genes. Other alternatives are already under development and should be available in the near future. Some early work with in vitro manipulations involving Tay-Sachs disease, a lethal genetic disease, has been demonstrated to be workable. The successful birth of a child without the genes for Tay-Sachs disease was accomplished even though both parents were carriers of the Tay-Sachs disease gene.

Another success was the test-tube fertilization to produce a child who would furnish critical cells to save a daughter who was dying from Franconi anemia - an incurable genetic disease. Both parents carried recessive genes for the Franconi anemia disease. Under in vitro manipulations eggs were taken from the mother and impregnated with sperm from the father. The fertilized eggs were examined for the gene carrying the disease. A disease free egg was located and then implanted into the mother, who eventually gave birth to a disease free baby boy. The unwholesome gene did not appear in the new offspring of the disease carrying parents.

Critical blood cells from the newborn boy were implanted into the dying daughter. The daughter's body responded to the new cells and is on the way to becoming a healthy adult. The baby boy is progressing normally. This test-tube fertilization again indicates that diseased genes can be successfully eliminated from being propagated into the next generation.

Evolution cannot function efficiently in a civilized society. Humankind takes care of the less able individuals. Sick or injured persons are not abandoned and left to the whims of fate, where predators might kill them. A blind person would have no chance to survive in a wild natural state. Yet in a civilized world they are secure and can live comfortably. There are few natural processes that will remove unfavorable genes in a civilized society. So a civilized society has to devise ways of eliminating unfavorable genes. If we do not, eventually, the human race could become inundated with unwholesome genes.

Individuals with unfavorable genes must make sure that their offspring do not carry the unfavorable genes into the future. That is the only way human genetic diseases can be eliminated. In the very near future our knowledge of genetic manipulation should become sophisticated enough to be put into action. In a few generations most of humankind's genetic diseases could disappear. It will require careful scrutiny and screening to accomplish the feat. But in the long run, the result would be a healthier, happier, and more productive population.

Imagine a world with no sickle cell anemia, no predisposition to breast cancer, no crippling rheumatoid arthritis, no predisposition to heart attack, no Alzheimer's disease, etc. It would make life so much more pleasant for us all. This is a desirable goal that can be reached. All that is needed is the wisdom and the will to achieve it. From a genetic health point of view, the future looks very bright.

Once again, religious ideas, social customs, racial ethics, etc., will impinge on any progress to eliminate genetic diseases. But despite the strident and persistent cries of foul, the good sense of eliminating genetic diseases will eventually prevail. The quaint, out-moded, and wrong ideas of many religions should not stand in the way of real progress in human health. Conquering genetic diseases will be accomplished, but it will take some reappraisal of humankind's needs and hopes to achieve it.

A final statement should be made about the ecology of the Earth. Every living thing that resides on the Earth has an influence on every other creature. Microbes influence our lives even though they may at first glance appear as insignificant. An essential function of our metabolism is escherichia coli in our intestines. Without these bacteria we would have difficulty surviving. The simple but broad statement, "that everything is connected to everything else," just about sums it up. Human beings are but one of the very large number of species of organisms that inhabit the Earth.

Now we are in a position to determine which types of various organisms will survive. It will be a formidable task to try to save all the myriad species that now exist in the world. Saving them will not be easy but we must strive to protect them all. Each one is unique. Each one is precious.

GLOSSARY

---

age of the solar system
The solar system formed about 4.6 billion years ago.

anthropology
The study of the origin, customs, characteristics, distribution, etc. of humankind.

archeology
The scientific study of material evidence such as: tools, buildings, etc., that remain from human cultures of the past.

artifact
Any object made by human skill or work.

asteroid
One of many thousands of relatively small, mostly rocky, planetlike objects that orbit about the sun.

background radiation or cosmic microwave radiation
A relic of the hot BIG BANG universe that now fills the present day universe.

biochemical
A chemical that is a part of the life process.

biochemists
Chemists that study the chemicals of life.

Burgess shale
Important source of exquisitely preserved fossils of early life forms.

Cambrian period of time
This geologic time zone covered about 505 million years from the present to about 550 million years ago.

catalyst
A substance that alters the speed of a chemical reaction without undergoing permanent change to itself.

comet
A relatively small body of ice and dust that orbits the sun in an exaggerated orbit.

continental drift
The movement of continents over the surface of the Earth.

cosmic rays
Atomic nuclei, mostly protons, that strike the Earth with very high velocities.

DNA
Deoxyribonucleic acid, the genetic material of life. Nuclear DNA is a huge polymer comprising over 3 billion base pairings. Mitochondrial DNA is still a big polymer but it only a tiny fraction of the size of nuclear DNA.

Edicaran fossils
These are the earliest macro-fossils ever located and are older than the Cambrian fossils. Their geologic time zone is in the range of about 550 million years ago to about 600 million years ago.

eukaryotic cell
A cell possessing a nuclei, mitochondria, chromosomes, and other complex internal structures.

evolution
A gradual process of change or development.

extraterrestrial life
Life process that developed independent of the Earth.

female circumcision or genital mutilation
The crude removal of the clitoris and other tissue from young girls.

Franconi Anemia
A fatal genetic disease inherited from parents.

galaxy
A huge assemblage of stars.

gene
A relatively short sequence of DNA that transmits hereditary characteristics.

glacier
A huge mass of ice that slowly moves over the Earth's surface.

half-life
The time required for half of a radioactive isotope sample to spontaneously decay.

Homo habilis
A hominid that produced the first stone tools about 2.5 million years ago.

hypothesis
A studied guess at explaining some phenomena.

in vitro manipulations
The use of laboratory equipment to bring about biological changes that normally take place in living systems.

isotopes
Atoms of the same element with different atomic weights.

light year
The distance that light can traverse in a vacuum in one year. It is more than 6 trillion miles.

meteorites
The surviving fragments of a relatively small objects from outer space that collided with the Earth.

Mid-Atlantic Ridge
The ridge on the Atlantic ocean floor where oceanic crust forms and then moves away in opposite directions.

Milky Way
The home galaxy of our solar system. On a clear night, far from city lights, a hazy band of light extends across the night sky. That is our view of the Milky Way from our position within the galaxy.

mitochondrial DNA
This relatively small DNA is found in the mitochondria of cells. Mitochondria provide energy for cells. Mitochondria DNA is inherited only from the mother.

mutant
An organism or individual that differs from the inherited characteristics of the parents.

mutation
A heritable change in the genes of an organism.

Neanderthal man
An extinct form of human that once flourished in Europe and some adjacent areas before disappearing about 35,000 years ago.

paleontologists
Scientists who study fossils.

Pompeii
A city in ancient Rome that was buried under volcanic debris in 79 AD.

pre-Cambrian
Any rocks older than Cambrian rocks are lumped together under pre-Cambrian. In terms of elapsed time the pre-Cambrian period occupies about 80% of total Earth time.

probabilities
The mathematics of random chance.

Proconsul
An early form of apelike creature that lived about 15 million years ago.

prokaryotic cell
A primitive cell that has no nuclei, chromosomes, mitochondria, etc., that characterize cells of higher organisms.

pseudo-protein
A polymer of the life system's amino acids that has a random orientation instead of the specific orientation of living proteins.

radioactive isotope
Atoms that are inherently unstable and will spontaneously undergo radioactive decay.

radioactivity
A process in which certain atomic nuclei spontaneously decompose and emit particles.

salinity
The concentration of dissolved salts in sea water.

San Andreas fault
The sites in California where the Pacific plate and the North American plate are rubbing together.

sedimentary rock
Rocks that have formed by slow sedimentation processes and then have solidified to become recognizable rocks.

sedimentation
The settling out of suspended particles from a fluid media such as dust in water.

self-catalyst
A molecular arrangement in which the fully formed molecule can make copies of itself, if the constituent parts are available.

solar system
The Sun and   its accompanying objects make up the solar systems. This includes the nine planets and their moons, the asteroids and comets, as well as all the fragments caused by random collisions.

speed of light
Light travels in a vacuum at 186,000 miles per second or 300,000 kilometers per second.

super-clusters of galaxies
Are clusters of clusters of galaxies encompassing many thousands of galaxies.

supernova
An extremely energetic explosion of a large star. Most of the mass of the star is blown away leaving a dense core.

Tay-Sachs
A lethal genetic disease inherited from both parents.

The BIG BANG
A super-colossal explosion-like event that initiated the universe about 13 to 14 billion years ago.

tides
The systematic movement of the oceans both up and down the beaches of the world.

topography
A precise and detailed description of an area of the world.

Tunguska event
Occurred in 1908 in Siberia near the headwaters of the Tunguska river. It was caused by an extraterrestrial object of moderate size striking th Earth. Trees were flattened over an elliptical region of about 25 x 30 miles or about 40 x 50 kilometers

INDEX

ABOUT THE AUTHOR

Dr. Firsching was born in Utica, N. Y. in 1923. He served for three years in the U. S. Army during WWII. He received an A. B. degree from Utica College and master's and doctoral degrees from Syracuse University. He worked for three companies as a research chemist before teaching at the University of Georgia for five years and Southern Illinois University at Edwardsville for 28 years.

He has 30 scientific publications. Included in these papers are original separation methods for barium and strontium and the rare earth elements, as well as extensive solubility data for the entire rare earth series. He maintained a weekly radio show and newspaper column, entitled "Think About It", for about 18 years.

He married Shirley Rae Gardner in 1954 and is the father of six children and the grandfather of nine. He has active hobbies in rock collecting, bird watching, and nature study. He has traveled in 49 states in the United States and took sabbatical leaves in California, Arizona, Colorado, and New York. He has visited Europe, South Africa, South Korea, and Brazil.

Some of the material for this book is taken from a course he developed called, "The Origins of Life", as well as assorted newspaper columns he had written.